Chocolate Wars

013496241 0

Chocolate Wars

From Cadbury to Kraft – 200 Years of
Sweet Success and Bitter Rivalry

DEBORAH CADBURY

Harper
Press

HarperPress
An imprint of HarperCollins*Publishers*
77–85 Fulham Palace Road,
Hammersmith, London W6 8JB
www.harpercollins.co.uk

Published by HarperPress in 2010

A catalogue record for this book is available from the British Library

ISBN 978-0-00-732555-9 (hb)
978-0-00-737485-4 (tpb)

Set in Goudy Old Style by Palimpsest Book Production Limited,
Falkirk, Stirlingshire

Printed and bound in Great Britain by Clays Ltd, St Ives plc

To Pete and Jo, Martin and Julia,
with love

Contents

Text Illustrations

Cadbury's cocoa works at Bridge Street, Birmingham, in the mid-nineteenth century. *(Cadbury Archives)*

George Cadbury in 1861. *(Cadbury Archives)*

Richard Cadbury in 1861. *(Cadbury Archives)*

John Cadbury's tea and cocoa shop in Bull Street, Birmingham, in 1824. *(Cadbury Archives)*

Candia at the time of her marriage to John Cadbury. *(Courtesy of John Crosfield)*

Dixon Hadaway. *(Cadbury Archives)*

A Cadbury's traveller on the road. *(Cadbury Archives)*

George Fox. *(The Religious Society of Friends in Britain)*

John and Candia Cadbury and their family in 1847. *(Cadbury Archives)*

The Birmingham workhouse. *(Courtesy of John Crosfield)*

The Rowntree factory at Tanners Moat in York in 1901. *(Reproduced from an original in the Rowntree-Mackintosh Archive at the Borthwick Institute, University of York)*

Fry's grinding machines. *(Cadbury Archives)*

Iceland Moss cocoa. *(Cadbury Archives)*

Van Houten's cocoa press. *(Cadbury Archives)*

George Cadbury in 1866. *(Courtesy of John Crosfield)*

Richard Cadbury with his son Barrow in 1866. *(Courtesy of John Crosfield)*

Cadbury's cocoa advertisement. *(Cadbury Archives)*

Horse-drawn omnibuses carried the Cadbury brothers' first poster campaign. *(Cadbury Archives)*

Richard's illustration of his daughter Jessie and her kitten. *(Cadbury Archives)*

Joseph Rowntree in 1862. *(Courtesy of the Joseph Rowntree Foundation)*

Daniel Peter's chocolate works at Vevey, Switzerland, c.1867. *(Nestlé Historical Archives, Vevey)*

Henri Nestlé, c.1875. *(Nestlé Historical Archives, Vevey)*

Rodolphe Lindt, c.1900. *(© Chocoladefabriken Lindt & Sprüngli AG)*

Nineteenth-century chocolate conching machine. *(Mary Evans Picture Library)*

The Bournville building site. *(Cadbury Archives)*

The first Bournville cottages were built around 1880. *(Cadbury Archives)*

The works forewoman. *(Cadbury Archives)*

Milton Hershey, c.1873. *(Courtesy of Hershey Community Archives, Hershey, PA)*

The Quaker schoolhouse Milton Hershey attended. *(Courtesy of Hershey Community Archives, Hershey, PA)*

Works steam train at Bournville. *(Cadbury Archives)*

Joseph Storrs Fry II. *(Cadbury Archives)*

George and Elsie Cadbury with their family. *(Courtesy of John Crosfield)*

Bournville village green. *(Cadbury Archives)*

Croquet in the women's recreation grounds. *(Cadbury Archives)*

The dining room at Bournville. *(Cadbury Archives)*

Milton and Kitty Hershey. *(Courtesy of Hershey Community Archives, Hershey, PA)*

The lily pond in the women's grounds at Bournville. *(Cadbury Archives)*

View of Bournville. *(Cadbury Archives)*

Ebenezer Howard's plan for a Garden City.

The Bournville dentist, 1905. *(Cadbury Archives)*

Drying cocoa in Trinidad, c.1896. *(Topfoto)*

In 1903 Milton Hershey began building his 'factory in a cornfield'. *(Courtesy of Hershey Community Archives, Hershey, PA)*

Bournville marzipan department, 1900. *(Cadbury Archives)*

Bournville's first laboratory. *(Cadbury Archives)*

Horse vans and 'Dennis' vans, 1906. *(Cadbury Archives)*

Milton Hershey with boys from the Hershey Industrial School, c.1923. *(Courtesy of Hershey Community Archives, Hershey, PA)*

George Cadbury. *(Courtesy of Adrian Cadbury)*

Cadbury's Easter egg. *(Cadbury Archives)*

Transporting milk to the Knighton factory in Staffordshire. *(Cadbury Archives)*

Laurence Cadbury. *(Courtesy of John Crosfield)*

Friends Ambulance Unit ambulance train. *(The Religious Society of Friends in Britain)*

Cadbury vans leaving the factory. *(Cadbury Archives)*

King George V and Queen Mary at Bournville in May 1919. *(Cadbury Archives)*

Bournville works outing, 1920. *(Cadbury Archives)*

The dining block at Bournville, 1927. *(Cadbury Archives)*

Women's gymnastics at Bournville, 1937. *(Cadbury Archives)*

Bournville camouflaged with netting during the Second World War. *(Cadbury Archives)*

Wartime first-aid class at Bournville. *(Cadbury Archives)*

A Cadbury 'cocoa caravan' brings comfort to a bombed-out city street. *(Cadbury Archives)*

Adrian and Dominic Cadbury. *(Courtesy of Adrian Cadbury)*

A Bournville works council. *(Cadbury Archives)*

A taste panel at Cadbury in the 1960s. *(Cadbury Archives)*

Cadbury India advertisement. *(Cadbury Archives)*

Adrian Cadbury. *(Cadbury Archives)*

Kraft cheese wagon, c.1921. *(Kraft®, Philadelphia®, Velveeta® and Miracle Whip® are registered trademarks of KF Holdings and are used with permission)*

Plates

Ripening cocoa pods. (*Eye Ubiquitous/Alamy*)

A chocolate pot depicted in Antonio de Pereda's *Still Life with an Ebony Chest* (1652). (*Hermitage, St Petersburg, Russia/The Bridgeman Art Library*)

The section on Trading from the *Christian and Brotherly Advices* (1738). (*Courtesy of the Library of the Society of Friends*)

A page from the 1801 *Extracts from the Minutes and Advices*. (*Courtesy of the Library of the Society of Friends*)

The Quaker Frys of Bristol had created the largest cocoa factory in the world by the mid-nineteenth century. (*Cadbury Archives*)

Fry's cocoa was selling all over England in the mid-nineteenth century, at a time when the Cadbury brothers' business appeared to be failing. (*Image courtesy of the Advertising Archives*)

The Cadburys' breakthrough product was a purer form of drinking cocoa, introduced in 1867. (*Cadbury Archives*)

Richard Cadbury's designs for the lids of 'fancy boxes' of chocolates first appeared in 1868. (*Cadbury Archives*)

Joseph Rowntree brought out a pure form of cocoa in the 1880s. (*Image courtesy of the Advertising Archives*)

George Cadbury's original sketch for a chocolate works at Bournville in 1878. (*Cadbury Archives*)

Joseph Rowntree built the model village of New Earswick near York in the early twentieth century. (*Borthwick Institute/HIP/TopFoto*)

Tennis for female staff at Joseph Rowntree's Haxby Road factory in 1900. (*Reproduced from an original in the Rowntree-Mackintosh Archive at the Borthwick Institute, University of York*)

Staff swimming lessons at Bournville, c.1910. (*Cadbury Archives*)

In the late nineteenth century Daniel Peter in Switzerland became the first manufacturer to create milk chocolate for eating. *(Nestlé Historical Archives, Vevey)*

Milton Hershey staked his fortune on chocolate in the early twentieth century. *(Courtesy of Hershey Community Archives, Hershey, PA)*

Rodolphe Lindt and Johann Sprüngli joined forces in 1899 to mass-produce chocolate from their new factory at Kilchberg, on the shores of Lake Zürich in Switzerland. *(C.H. Kilchberg © Chocoladefabriken Lindt & Sprüngli AG)*

The American Forrest Mars made his first Mars Bar in Slough in 1933, and sold two million bars in a year. *(Image courtesy of the Advertising Archives)*

1

Introduction

When I was a young child, the knowledge that a branch of my family had built a chocolate factory filled me with wonder. What sort of charmed life did such a possibility offer to my relatives? Each Christmas I had an insight when the most enormous case arrived from my uncle, Michael Cadbury, containing a large supply of mouth-watering chocolates. Even more memorable was the trip I made in the early 1960s to see how the chocolate was made. As I opened the door to the factory at Bournville in Birmingham, the sight that greeted me was magical.

To my child's eyes it was as though I had entered a cavernous interior that belonged to some benign, orderly and highly productive wizard who had somehow saturated the very air with a chocolate aroma. My uncle and parents raised their voices against the whirr of machinery. But I did not hear them. All I could see was chocolate. It was all around me, in every stage of the manufacturing process. There was molten chocolate bubbling in vats towering above me, vats so huge that they had ladders running up their sides. Chocolate rivers flowed on a number of swiftly-moving conveyers through gaps in the wall to mysterious chambers beyond. Solid chocolate shaped in a myriad of exciting confections travelled in neat, soldierly processions towards the wrapping department. Such a miracle of clockwork precision and sensual extravagance was hard to take in. Even more puzzling to my young mind was the question of how this chocolate feast, which brought the idea of greed to a whole new level, fitted with religion? For even though I did not yet understand the connection, I did know that the chocolate works were, in some inexplicable way, intimately connected with a religious movement known as Quakerism. Was all this the hand of God?

My own father had left the Quakers just before the Second World War. He wanted, as he put it, to 'join the fight against Hitler', a stance that was not compatible with Quaker pacifism. I was brought up in the Church of England, and as a child, when I joined my cousins for Quaker meetings, I felt as if I were on the outside looking in on a strange, even mystical tradition. Long silences endured in bare rooms, stripped of anything that might excite the senses, where grownups contemplated the surrounding void, were incomprehensible to me. Equally incomprehensible: how did my rich, chocolate relatives acquire that admirable restraint, that air of wholesome frugality? Even family picnics had a way of turning into long and chilly route marches, raindrops trickling down my back. The wealth and the austerity seemed oddly incongruous. Did the one contribute to the other? Cheerful homilies from my father along the lines of 'Many a mickle makes a muckle' and 'Look after the pennies and the pounds will look after themselves' did not supply a satisfactory answer. Even a five-year-old knew this was not the key to creating a chocolate factory.

A generation passed before I decided to retrace my steps up Bournville Lane. This time it was personal. I wanted to delve into the Bournville and family archives to uncover the whole story. When I turned the corner in the lane in the autumn of 2007, my heart skipped a beat as I was taken back to that day when my father and uncle, both now much missed, had taken me round the factory. To my surprise, the chocolate works seemed even larger than I remembered. Imposing red-brick blocks stood beside the neatly mowed cricket pitch, with Bournville village and green nestled behind. At this time, Cadbury was the largest confectioner in the world, and the only independent British chocolate enterprise to survive from the nineteenth century. I wanted to understand the journey that took my deeply religious Quaker forebears from peddling tins of cocoa from a pony and trap around Birmingham to this mighty company that reached around the globe.

The story began five generations ago, when the far-sighted Richard Tapper Cadbury, a draper in Birmingham in the early nineteenth century, sent his youngest son, John, to London to study a new tropical commodity that was attracting interest among the colonial brokers of

Mincing Lane: cocoa. Was it something to eat or drink? Richard Tapper saw it pre-eminently as a nutritious non-alcoholic drink in a world that relied on gin to wash away its troubles. Never could my abstemiously inclined ancestor have guessed what fortunes would be entwined with the humble cocoa bean, although it seemed full of promise, a touch of the exotic.

His grandsons, George and Richard Cadbury, turned a struggling business into a chocolate empire in one generation. In the process, they found themselves in competition with their Quaker friends and rivals Joseph Rowntree in York, and Francis Fry and his nephew Joseph in Bristol. The Cadbury, Fry and Rowntree dynasties were built on values that form a striking contrast with business ethics today. Their approach to the creation of wealth was governed by a code of practice developed over hundreds of years since the English Civil War by their Quaker elders and set out at yearly meetings and in Quaker books of discipline. This nineteenth-century 'Quaker capitalism' was far removed from the excesses of the world's recent financial crisis, with business leaders apparently seeing no harm in pocketing huge personal profits while their companies collapsed.

For the Quaker capitalists of the nineteenth century, the idea that wealth-creation was for personal gain only would have been offensive. Wealth-creation was for the benefit of the workers, the local community and society at large, as well as for the entrepreneurs themselves. Reckless or irresponsible debt was also seen as shameful. Quaker directives ensured that no man should 'launch into trading and worldly business beyond what they can manage honourably . . . so that they can keep their words with all men'. Even advertising was dismissed as dishonest, mere 'puffery': the quality of the product mattered far more than the message. Men like Joseph Rowntree and George Cadbury built chocolate empires at the same time as writing ground-breaking papers on poverty or studies of the Bible, or campaigning against a multitude of Dickensian human rights abuses. Puritanical hard work and sober austerity, with the senses kept in watchful restraint, were the guiding principles.

While it is easy to dismiss such values as antiquated, Quaker capitalism proved extraordinarily successful, and generated a staggering amount of worldly wealth. In the early nineteenth century, around 4,000 Quaker families in Britain ran seventy-four banks and over two hundred companies. As they came to grips with making money, these austere men of God also helped to shape the course of the Industrial Revolution, and the commercial world today.

The chocolate factories of George and Richard Cadbury and Joseph Rowntree inspired men in America such as Milton Hershey, the 'King of Caramel', who took philanthropy to a new, all-American scale with the creation of the utopian town of Hershey in the cornfields of Pennsylvania. But the chocolate wars that followed the growth of global trade, and the emergence in the twentieth century of international rivals – such as Frank and Forrest Mars – unshackled by religious conviction, gradually eroded the values that had shaped Quaker capitalism. Some Quaker firms did not survive the struggle, and those that did had little choice but to abandon their Puritan roots. In the process, ownership of the businesses passed from private Quaker dynasties to public shareholders. Little by little, the results of the transition from Quaker capitalism to shareholder capitalism began to take shape in the form of the huge confectionery conglomerates that straddle the corporate world.

Today the world's two largest food giants – the Swiss Nestlé and America's Kraft – operate around the globe, feeding humanity's sweet tooth. Nestlé, with five hundred factories worldwide, sells a billion products every day, giving annual sales of £72 billion. Kraft operates 168 factories, and has annual sales of £26 billion. While these two behemoths are locked in a race to maintain market share in the developed world, they are also selling their Western confections and other processed foods to emerging markets in the developing world. Somewhere along the way the four-hundred-year-old English Puritanical ideal of self-denial and the Quaker vision of creating wholesome nourishment for a hungry and impoverished workforce have disappeared. Also vanished are a myriad of independent chocolate confectionery firms. In Britain alone, Mackintosh of Halifax and Rowntree of York are now owned by Nestlé, while Terry of York, Fry of Bristol and Cadbury have become a division of Kraft.

The origins of this book lie in my search to explore how this happened. I wanted to unearth the true story of the original Quaker chocolate pioneers and the religious beliefs that shaped their business decisions, and to see how their values differ from those of today's company leaders. At first sight it can appear that globalisation has been profitable for all. It is hard to dispute economists' claims that the process has lifted billions of people across the world out of the poverty that was on the doorstep of the cocoa magnates of the nineteenth century. But that has also come at a significant cost.

My last visit to Bournville, on a bitterly cold January day in 2010, formed a stark contrast to the peaceful charm of my earlier visits. Outside the factory, staff members with banners were protesting against Kraft's recently announced takeover of Cadbury. 'Kraft go to hell,' said one. In a symbolic gesture, another protester set fire to a huge Kraft Toblerone bar. Unite, Britain's biggest trade union, had warned that thousands of jobs could be eliminated under Kraft. 'Our members feel very angry and very betrayed,' said Jennie Formby, Unite's national officer for food and drink industries. Kraft was borrowing £7 billion to fund the takeover, and many feared that Cadbury could become 'nothing more than a workhorse used to pay off this debt', its assets stripped and jobs lost. For the Quaker pioneers, the workforce and the local community were key stakeholders in the business, and they aimed to enhance their lives. Now, with their future uncertain, there was a mood of alienation and powerlessness amongst the staff outside Bournville that day.

Quite apart from the effects on employees and communities, there are additional concerns raised by the Kraft takeover of Cadbury that bring the contrast between Quaker business values and shareholder capitalism sharply into focus. For the nineteenth-century Quaker, ownership of a business came with a deep sense of responsibility and accountability to all stakeholders involved. In today's system of shareholder capitalism the shareholder is divorced from the responsibilities of business ownership.

The spirit of a business – so crucial to the motivation of its staff – is hard to define or measure. It is not to be found in the buildings or the balance sheet, but it is reflected in the myriad of different decisions taken by those at the helm of the business. The Quaker pioneers believed that 'your own soul lived or perished according to its use of the gift of life'. For them, spiritual wealth rather than the accumulation of possessions was the 'enlarging force' that informed business decisions. But gone now, lost in another century, is that omnipotent all-seeing eye in the boardroom, reminding those Quaker patriarchs of the fleeting nature of their power. And what is there to replace it?

The story of the Quaker chocolate pioneers and their rivals is, in a way, a parable of our times, highlighting a bigger transformation in our society. By examining the 'chocolate wars' that have shaped the world of confectionery, I hope to shed light on a process of change that affects us all.

PART ONE

1

A Nation of Shopkeepers

In the mid-nineteenth century, Birmingham was growing fast, devouring the surrounding villages, woods and fields. The unstoppable engine of the Industrial Revolution had turned this once modest market town into a great sprawling metropolis in the heart of the Midlands. Country-dwellers hungry for work drove the population from 11,000 in 1720 to more than 200,000 by 1850. In the city they found towering chimneys that turned the skies thunder grey, and taskmasters unbending in their demands. Machines never stopped issuing the unspoken command: more toil to feed the looms, to fire the furnaces and to drive the relentless wheels of commerce and industry far beyond English shores.

Birmingham was renowned across the country for innovation and invention. According to the reporter Walter White, writing about a visit to the city in *Chambers's Edinburgh Journal* in October 1852, 'To walk from factory to factory, workshop to workshop and view the extraordinary mechanical contrivances and ingenious adaptations of means to ends produces an impress upon the mind of no common character.' The town was a beacon of industrial might and muscle. This was where fire forged iron and coke, metal and clay to make miracles.

Birmingham's foggy streets resounded with hammers and anvils fashioning bronze and iron into buttons, guns, coins, jewellery, buckles and a host of other Victorian artefacts. Walter White marvelled at the 'huge smoky toyshop' and the 'eager spirit of application manifested by the busy population'. But he was evidently less taken with the sprawling town itself, which he considered 'very ill arranged and ugly', and dismissed as 'a spectacle of dismal streets'.

At the heart of these dismal streets, opposite today's smartly paved Centenary Square, was a road called Bridge Street, which in 1861 was the site of a Victorian novelty: a cocoa works. Approached down a dirt road, past busy stables, coach houses and factories, it was surprisingly well hidden. But wafting through the grimy back-streets was a powerful aroma, redolent of rich living. Guided by this heady perfume the visitor was drawn past the blackened exterior, through a narrow archway into a courtyard with an entrance leading off to the heart of the chocolate factory. It was to this modest retreat that two young Cadbury brothers hurried one day early in 1861.

There was a crisis in the family. Twenty-five-year-old Richard and twenty-one-year-old George Cadbury knew that the wonderful aroma of chocolate disguised a harsher reality. The chocolate factory and its owner, their father, John Cadbury, were in decline. The family faced a turning point. The business could go under completely. John Cadbury turned to his sons for help.

Photographs of the time show George and Richard Cadbury soberly dressed in plain dark Victorian suits with crisp white shirts and bow ties. Richard's soft features contrast with his younger brother, whose intensity of focus and air of concentration is not relaxed even for the photographer. 'I fixed my eye on those who had won,' George admitted later. 'It was no use studying failure.' He had, said his friends, 'boundless ambition'. And he needed it. The family firm was haemorrhaging money.

By chance, Walter White toured the Bridge Street factory, and has left a vivid account of what it was like in 1852. Leaving behind the storehouse crammed with sacks of raw cocoa beans from the Caribbean, he entered a room that blazed with heat and noise: the roasting chamber. With its four vast rotating ovens, 'the prime mover in this comfortable process of roasting was a 20 horse steam engine'.

After this, 'with a few turns of the whizzing apparatus', the husk was removed by the 'ceaseless blast from a furious fan' and the cocoa, 'now with a very tempting appearance', was taken for more 'intimate treatment'. This occurred in a room where 'shafts, wheels and straps kept a number of strange looking machines in busy movement'. Following yet more pressing and pounding, finally a rich, frothing chocolate mixture flowed, 'leisurely like a stream of half frozen

Cadbury's cocoa works at Bridge Street, Birmingham, in the mid-nineteenth century.

treacle'. This was formed into a rich cocoa cake which was shaved to a coarse powder ready for mixing with liquid for drinking. Upstairs, White found himself in a room where management 'had put on its pleasantest expression'. The female employees, all dressed in clean white Holland pinafores, were 'packing as busily as hands could work. No girl is employed,' he added, 'who is not of a known good moral character.' Such a factory, he concluded, was 'a school of morality and industry'.

But almost ten years had passed since Walter White's visit, and the 'school of morality and industry' had been quietly dying of neglect. George Cadbury was quick to appraise the desperate situation. 'Only eleven girls were now employed. The consumption of raw cocoa was so small that what we now have on the premises would have lasted about 300 years,' he wrote. 'The business was rapidly vanishing.'

During the spring of 1861, George and Richard wrestled with their options. Pacing the length of the roasting room in the evenings, the four giant rotating ovens motionless, the dying embers of the coke fire beneath them faintly glowing, the brothers could see no simple solution. George had harboured hopes of developing a career as a doctor. Should he now join his brother in the battle to save the family chocolate business? Or close the factory? Would they be able to succeed, where their own father had failed?

It was their father, John, who had proudly shown Walter White around the factory in 1852. In the intervening years he had been almost completely broken by the death of his wife, Candia. John had watched her struggle against consumption for several years, her small frame helpless against the onslaught of micro-organisms unknown to Victorian science. Equipped only with prayers and willpower, he took her to the coast hoping the fresh air would revive her, and brought in the best doctors.

But nothing could save her. By 1854, Candia had gratefully succumbed to her bath chair. Eventually she was unable to leave home, and then her room. 'The last few months she was indeed sweet and precious,' wrote John helplessly. When the end came in March 1855, 'Death was robbed of all terror for her,' he told his children. 'It was swallowed up in victory and her last moments were sweet repose.' Yet as the weeks following her death turned into months John failed to recover from his overwhelming loss. He was afflicted by a painful and disabling form of arthritis, and took long trips away from home in search of a cure. After years of diminishing interest in his cocoa business, Cadbury's products deteriorated, its workforce declined and its reputation suffered.

Richard and George knew their father's cocoa works was the smallest of some thirty manufacturers that were trying to develop a market in England for the exotic New World commodity. No one had yet uncovered the key to making a fortune from the little bean imported from the New World. There was no concept of mass-produced chocolate confectionery. In the mid-nineteenth century, the cocoa bean was almost invariably consumed as a drink. Since there was no easy way of separating the fatty cocoa oils, which made up to 50 per cent of the bean, from the rest, it was visibly oily, the fats rising to the surface. Indeed, it often seemed that the novelty of purchasing this strange product was more thrilling than drinking it.

John Cadbury, like his rivals, followed the established convention of mixing cocoa with starchy ingredients to absorb the cocoa butter. As his business had declined, the proportion of these cheaper materials had increased. 'At the time we made a cocoa drink of which we were not very proud,' recalled George Cadbury. 'Only one fifth of it was cocoa – the rest was potato flour or sago and treacle: a comforting gruel.'

This 'gruel' was sold to the public under names such as Cocoa Paste, Soluble Chocolate Powder, Best Chocolate Powder, Fine Crown, Best Plain, Plain, Rock Cocoa, Penny Chocolate and Penny Soluble Chocolate. Customers did not buy it in the form of a powder but as a fatty paste, made up into a block or cake. To make a drink at home, they chipped or flaked bits off the block into a cup and added hot water, or if they could afford it, milk. It is a measure of how badly the Cadbury cocoa business was faring that three-quarters of the Bridge Street factory's trade came from tea and coffee sales.

Although promoted as a health drink, cocoa had a mixed reputation. Unscrupulous traders sometimes coloured it with brick dust and added other questionable products not entirely without problems for the digestive system: a pigment called umber, iron filings or even poisons like vermilion and red lead. Such dishonest dealers also found that the expensive cocoa butter could be stretched a little further with the addition of olive or almond oil, or even animal fats such as veal. The unwary customer could find himself purchasing a drink which could not only turn rancid, but was actually harmful.

While the prospects for the family business in 1861 did not look hopeful, the alternatives for Richard and George were limited. Quakers, like all non-conformists, were legally banned from Oxford and Cambridge, the only teaching universities in England at the time. As pacifists, they could not join the armed services. Nor were they permitted to stand as Members of Parliament, and they faced restrictions in other professions such as the law. As a result, many Quakers turned to the world of business, but here too the Society of Friends laid down strict guidelines.

In a Quaker community, a struggling business was a liability. Failing to honour a business agreement or falling into debt was seen as a form of theft, and was punished severely. If the cocoa works went under owing money Richard and George would face the censure of the Quaker movement; at the worst, they could be disowned completely and treated as outcasts. Quite apart from these strict Quaker rules, in Victorian society business failure and bankruptcy could lead to the debtors' prison or the dreaded workhouse, either of which could lead to an early grave.

Ahead was a battle in which defeat was all too possible. The brothers did not have to dedicate themselves to this struggling enterprise, with its cramped premises in which their offices were scarcely bigger than coffins. 'It would have been far easier to start a new business, than to pull up a decayed one which had a bad name,' George said later, looking back on his life. 'The prospect seemed a hopeless one, but we were young and full of energy.'

George Cadbury in 1861, aged twenty-one, at the time that he took over the failing factory at Bridge Street.

Richard Cadbury in 1861, aged twenty-six.

To the remaining employees of the company, who now had reason to fear for their jobs, 'Mr Richard' appeared jovial, relaxed and 'always smiling', while 'Mr George' was cut from a different cloth, 'stern but very just'. His unremitting self-discipline and his ability to focus every aspect of his life on one goal became legendary. 'He was not a man,' a colleague later observed, 'but a purpose.' And what George and Richard decided to do next would become the stuff of family legend.

Richard and George Cadbury were the third generation of Cadbury tradesmen in Birmingham. It was their grandfather, Richard Tapper Cadbury, who had been instrumental in breaking centuries of long association with the West Country and leading the family in a new direction as shopkeepers in the town. At the close of the eighteenth century, as Napoleon prepared for his long march over Europe, the Cadburys, like countless others at the time, exemplified the Britain that the French leader dismissed as a mere 'nation of shopkeepers'. And just as Napoleon's scathing remark underestimated his enemy's real wealth and capacity for war, so it was easy not to see the huge potential emerging from a new generation of shopkeepers whose connections were only just beginning to reach out across the world.

Like many Victorians, the young Richard Cadbury had a fascination with family history, and when travelling he took the opportunity to study 'the records of his ancestors with thoroughness and affection'. Eventually he compiled a beautifully illustrated *Family Book*. This shows how his forebears lived as sheep farmers and wool combers in the West Country, their quiet lives marked for centuries by nothing more dramatic than the passing seasons, until 1782 when Richard and George's grandfather, Richard Tapper Cadbury, set out to learn a trade.

'Very little is known of his early life,' writes Richard of his grand-father. 'He left home in Exeter when he was fourteen on the top of the coach . . . to serve as an apprentice to a draper.' The young Richard Tapper remembered the morning of his leaving: 'My father and mother got up early to see me off by the stage . . . and I thought my heart would break.'

Richard Tapper was apprenticed to a draper 150 miles away in Kent who supplied uniforms to troops fighting in the American War of Independence. Within a year the war ended, the troops were demo-bilised and the business went bankrupt. Richard Tapper secured another position as an apprentice in Gloucester, where by the age of nineteen he was proud to receive wages of £20 a year. After 'scrupu-lously and conscientiously' avoiding any 'unnecessary gratification', he reassured his parents in Exeter, it was possible for him to pay for his own washing and 'appear so respectable as to be invited as guest among the first families of Gloucester'. His next move was to London, to work for a linen draper and silk mercer in Gracechurch Street. His wages eventually rose to £40 a year, which not only enabled him to 'maintain a respectable appearance' but also to 'purchase many books'.

After ten long years mastering the trade, Richard Tapper was longing to start a draper's business of his own. He was dissuaded from his youthful dream of sailing for America by a family friend, who warned him 'that the country is still far from settled'. Nor could he seek adventures in Europe, with France in the frenzied grip of Robespierre's Reign of Terror and at war with its neighbours. So in 1794, equipped with much enthusiasm and, through the Quaker network, quite a few references, he took the stage to Birmingham with a friend, Joseph Rutter. They had heard of an opening for a linen draper and silk mercer in the town, and seized their chance.

The draper's shop at 92 Bull Street was soon successful enough for Richard Tapper to buy out his partner and start a family. He married Elizabeth Head in 1796, and seven children followed over the next seven years. Elizabeth still found time to help in the shop, dressing the windows with fine silks and linens and taking an interest in the changing fashions. One year they were obliged to enlarge their front door to accommodate the fashion for puffed 'gigot' sleeves, strengthened with feather pads or whalebone hoops. Records show that Richard Tapper's business was so successful that in 1816 a second shop at 85 Bull Street was also registered in his name.

The stories that survive in Richard Cadbury's *Family Book* provide a vivid glimpse into his grandfather's life. One of the problems Richard Tapper had to deal with was theft. After repeatedly losing silk that cost up to twelve shillings a yard, he felt he had to take action, but soon came to regret it. He stopped a woman in his shop who had two rolls of silk hidden under her cloak. When he went to court to hear the outcome of her trial, to his alarm the judge sentenced the woman to death. 'I was appalled,' Richard Tapper told his children years later, 'for I never realised what the sentence would be. Without delay I posted to London, saw the Secretary of State and got the woman's sentence commuted to transportation.' Given that this was a time when it took almost two days by 'Flying Coach' – with fresh horses staged down the line – to reach London from Birmingham, this required considerable commitment.

As a Quaker, Richard Tapper became deeply involved in community affairs, and served on the Board of Street Commissioners for Birmingham, a precursor to the Town Council. He also worked as an Overseer of the Poor, including during the troubled year of 1800, when the harvest failed. According to the *St James Chronicle*, on 8

October the price of bread rose to nearly two shillings for one loaf. In the Parish of Birmingham the poor were in dread of starvation, 'the distress in the town was great', and there was 'an alarming disorder' in the workhouse. Richard Tapper was among those who tried to ensure that there was enough food.

Richard Tapper's business prospered, and his garden at the back of 92 Bull Street became a favourite spot for his growing family, with 'currants in abundance, flowers and a vine'. The accounts of Richard's children are of particular personal interest since my own branch of the family can be traced to his oldest son, Benjamin, born in 1798. According to the *Birmingham Daily Post*, Benjamin had a passion for philanthropy. Among the many benevolent causes he supported were the local Infant Schools, the Bible Society and the Society for the Suppression of Cruelty to Animals. But like many Quakers, stated the *Post*, by far his 'most laborious and anxious labours' were devoted to the anti-slavery movement, 'which more or less occupied his time and unwearied attention for upwards of thirty-five years'. When Benjamin turned thirty he inherited his father's successful draper's shop on Bull Street, and was happily settled there for many years.

Richard Tapper Cadbury's second son, Joel, was able to fulfil his father's own dream of seeking his fortune in America, and set sail in 1815 at the age of sixteen. The stormy Atlantic crossing took eighty days in high winds and rough seas that washed a man overboard and prompted seasoned sailors to say they had never seen such a sea. Joel eventually settled in Philadelphia, and became a cotton goods manufacturer. He had a family of eleven children, and established a large branch of Quakers on the east coast of America.

Richard's third son, John – the father of Richard and George – born in 1801 above his father's draper's shop, was destined to have a very different fate. According to an account handed down through generations, John's far-sighted father, having passed on his business to his oldest son, Benjamin, asked John to investigate the new colonial market in Mincing Lane, London. He was curious about a new commodity, the cocoa bean, which was arriving from the New World.

Today, among the gleaming black façades of City office blocks, there is little to give away Mincing Lane's colourful past. The only hint of its nineteenth-century purpose survives in the name of a building halfway

down the street: Plantation House. But when John Cadbury visited in
the 1820s, it was a teeming market where colonial brokers met to trade
in commodities from Britain's growing empire. There were sale rooms
where frenetic auctions were taking place for tea, sugar, coffee, jute,
gums, waxes, vegetable oils, spices and cocoa. Prices and details of business were written on a blackboard, and samples of goods from warehouses
in the docks along the nearby Thames were on display. They included
the cocoa bean, or 'nib', from South America, which looked like a huge
chocolate-coloured almond, still dusted with the dried pulp that
surrounded it in the cocoa pod, and baked by the tropical sun.

At a time when cocoa was purchased primarily to produce a novelty
drink for the rich, John's mission was to ascertain whether there might
be a future in this unpromising black bean.

John Cadbury, like his father before him, had set out to learn his
trade at a young age as an apprentice. In 1816, aged fifteen and
proudly dressed in the best-quality cloth from the family's draper's
shop, he took the coach journey to Leeds, where he was apprenticed
to a Quaker tea dealer. It appears he made a good impression. His
aunt, Sarah Cash, who visited the following year, declared, 'John is
grown a fine youth, he possesses a fine open countenance, is vigorous
of mind and body and desires to render himself useful.' Others
commented that the plain Quaker boy, soberly dressed in dark colours,
stood out next to the rough Yorkshire boys. It seems the owners were
soon content to leave the care of their tea business with John when
they had to travel, and he was rewarded on his departure after seven
years with a fine encyclopaedia.

John went to London to be apprenticed at the teahouse of
Sanderson Fox and Company. While in London he went to see the
warehouses of the East India Company, and watched the sale of
commodities such as coffee and cocoa. He wrote to his father that
he was convinced that there was potential in the exotic new bean,
although he was not yet clear what that potential was.

In 1824 the twenty-three-year-old John returned to Birmingham
and set up a tea and coffee shop of his own on Bull Street, next door
to his brother Benjamin's draper's shop. His father lent him a small

sum of money and said he must 'sink or swim', as there were no further funds. John proudly announced the opening of his shop in the local paper, Aris's *Birmingham Gazette*, on 1 March. After setting out his considerable experience 'examining the teas in the East India Company's warehouses in London', he drew the public's attention to something new: a substance 'affording a most nutritious beverage for breakfast . . . Cocoa Nibs prepared by himself'.

John Cadbury's tea and cocoa shop in Bull Street, Birmingham, in 1824.

John Cadbury took advantage of the latest ideas to draw business to his shop, starting with the window. While most other shops had green-ribbed windows, John's had innumerable small squares of plate glass, each set in a mahogany frame, which it was said he polished himself each morning. This alone was such a novelty that people would come to see it from miles around. On peering through the glass, prospective customers would be intrigued to see a touch of the Orient in the heart of smoky Birmingham. The many inviting varieties of tea, coffee and chocolate were displayed amongst handsome blue Chinese vases, Asian figurines and ornamental chests. Weaving his way through all this exotica was a Chinese worker in Oriental dress. Those who ventured inside were greeted by the rich aroma of coffee and chocolate; John ground the cocoa beans himself in the back of the shop with mortar and pestle. Word of John Cadbury's quality teas and coffees soon spread amongst the wealthiest and best-known families in Birmingham: his customers included the Lloyds, Boultons, Watts, Galtons and others.

Meanwhile, through the Quaker network, John met Candia Barrow of Lancaster. The Barrows and the Cadburys had already developed very close ties through marriage. In 1823 John's older sister Sarah had married Candia Barrow's older brother. This was followed in 1829 by the marriage of John's older brother Benjamin to Candia's cousin, Candia Wadkin. So in June 1832, when John married Candia Barrow, it was the third marriage in a generation to link the two Quaker families. It proved to be a very happy union.

As John's shop prospered, he could see for himself the growing demand for cocoa nibs. He took advantage of the large cellars under the shop to experiment with different recipes, and created several successful cocoa powder drinks. So confident was he of the future of this nutritious and wholesome drink that he decided to take a further step: into manufacturing.

In 1831, John rented a four-storey building in Crooked Lane, a winding back street at the bottom of Bull Street, and began to test produce cocoa on a larger scale. The idea of using machines to help process food was in its infancy, but to help him with the roasting and pressing of beans he installed a steam engine, which evidently was a great family novelty. According to his admiring aunt Sarah Cash, everyone in the family 'had thoroughly seen John's steam engine'.

Candia at the time of her marriage to John Cadbury.

After ten years he had developed a wide variety of different types of cocoa for his shop: flakes, powders, cakes, and even the roasted and crushed nibs themselves.

Meanwhile, Candia and John started a family, and moved to a house with a garden in the rural district of Edgbaston. Their first son, John, suffered intermittently from poor health. Richard Cadbury, their second child, was born on 29 August 1835, and was followed by a sister, Maria, and then George, born on 19 November 1839. The arrival of two more sons, Edward and Henry, would complete the family.

To the boys' delight their parents placed a strong emphasis on the enjoyment of an outdoor life. Their house had a square lawn, recalled Maria. 'Our father measured it round, 21 times for a mile, where we used to run, one after another, with our hoops before breakfast, seldom letting them drop before reaching the mile, and sometimes a mile and a half, which Richard generally did.' Only then were they allowed in for breakfast, 'basins of milk . . . with delicious cream on top and toast to dip in'. After this early-morning ritual, John set off to work. 'I can picture his rosy countenance full of vigour,' says Maria, 'his Quaker dress very neat with its clean white cravat.'

A memorable delight for the boys was the arrival of the railway in

Birmingham. Britain was in the grip of railway fever, and the Grand Junction Railway steamed into Birmingham from Manchester in 1837. Within a year, a line opened that covered the hundred miles between Birmingham and London. The treacherous two-day journey to the capital by horse and coach was reduced to a mere two hours by steam train. The coming of the railway made a deep impression on the growing family. 'I got a railway train, first second and third class carriages, with an engine and tender,' seven-year-old Richard wrote enthusiastically to his brother in 1842. For his father, however, it opened up whole new possibilities for trade.

At the age of eight, Richard was sent away to join his older brother John at boarding school. George studied with a local tutor who had a decidedly individualistic view on the best way to deal with boys. He aimed to instil mental and physical fortitude with a diet of classics and combative sports, including occasional games of 'Attack' which he devised himself, and which involved arming the boys with sticks. Somehow George came through the experience with a sound knowledge of French and Virgil, and a keen appreciation of home life. He remembered his childhood as 'severe but happy', with an emphasis on discipline and a lifestyle that was 'bare of all self indulgence and luxury'.

Both George and Richard formed vivid impressions of trips to see their mother's family at Lancaster. Their maternal grandfather, George Barrow, in addition to running a draper's shop, had created a prosperous shipping business with trade to the West Indies. His grandchildren were allowed to climb the tower he had built in the grounds of his house, from where they had a stunning view of Morecambe Bay and on occasion his returning ships, sometimes banked up three at a time on the quayside beyond. Sea captains came to visit, and would regale the children with tales of wide seas and foreign lands, the wonders of travel and the horrors of the slave trade.

By 1847 John Cadbury's Crooked Lane warehouse had been demolished to make way for the new Great Western Railway. Undeterred, John expanded his manufacturing into the Bridge Street premises, and was soon joined by his older brother Benjamin. By 1852 the two brothers were in a position to open an office in London, and they later received a royal accolade as cocoa manufacturers to Queen Victoria. It was around this time that they dreamed up a plan to create a model village for their workers, away from the grime of the city; they even designated one of their brands of cocoa 'The Model Parish Cocoa'.

In 1850, when he was almost fifteen, Richard joined his father and uncle at Bridge Street, and was doubtless aware of their grand ambitions. With his father often away, he threw himself into the trade with 'energy and devotion', observed one relative. Richard kept his father informed of day-to-day events: 'James is very steady at the engine, keeps it at a regular pace and in beautiful order and is careful not to *waste* any money over it. The girls do their work cheerfully, but want a good deal of looking after . . . ' .

The pressures of learning the trade did not stop Richard indulging his love of sport. He and George were passionately fond of skating, and would often rise at 5 a.m. so as to be on the ice before dawn. 'Only those who have made the effort know the exhilaration of skating in the early morning and watching the light gradually break and the beauty of the sunrise,' wrote George. One young friend of his sister Maria remembered that Richard 'used to fairly dazzle us with his skating'. But events were conspiring against such relaxed pursuits.

Cocoa sales had begun to decline during the economic depression of the 'Hungry Forties', when a slump in trade, rising unemployment, bad harvests and a potato blight in England and Scotland in 1845 combined to create widespread hardship. Many small businesses struggled, but for the Cadburys the irrevocable blow came in the early 1850s, when Candia was diagnosed with tuberculosis.

These painful years left their mark on Richard and George. They witnessed the inexorable decline first of their mother then of their father, then the neglect of the factory, as though it too was afflicted with a malady for which there was no known cure. John still occasionally walked through the factory in his starched white collar and neat black ribbon tied in a bow, but the enthusiasm that had prompted him to develop the venture over a period of thirty years was gone. He paid scant attention to the piles of cocoa beans accumulating in the stock room. The hard-won accolade as cocoa manufacturer to Queen Victoria no longer excited him. A year after Candia's death he dissolved the partnership with his brother Benjamin. Gradually his absences became more prolonged as he searched for a cure for his arthritis, and the family firm began to lose its good name.

These were the pressing concerns in young George Cadbury's mind when in 1857, like his father and grandfather before him, he too was sent away to learn his trade as an apprentice. His sister Maria had taken his mother's place in the home looking after the younger children. His older brother Richard was taking on more responsibility for his father's business. George was keen to master the trade by working in a grocery shop in York run by another Quaker, Joseph Rowntree.

Once past York's famous city walls, the seventeen-year-old George Cadbury found himself in a maze of winding old streets, with irregular gabled houses, the overhang of their upper storeys making the streets narrow and dark. At the bottom of the Shambles, the road opened onto a busy thoroughfare called Pavement. Almost directly opposite, at number 28, stood Rowntree's shop, a handsome eighteenth-century terraced house, tall and narrow, its walls made crooked by subsidence. There was little in the colourful thoroughfare outside to hint at the austerity and long hours that awaited George inside the shop.

The strict rules of conduct that Joseph Rowntree expected his numerous apprentices to obey were clearly set out in a *Memorandum*:

> *The object of the Pavement establishment is business. The young men who enter it as journey men or apprentices are expected to contribute . . . in making it successful . . . It affords . . . a full opportunity for any painstaking, intelligent young man to obtain a good practical acquaintance with the tea and grocery trades . . . The place is not suitable for the indolent and the wayward . . .*

The *Memorandum* specified every detail of the boys' lives: no more than twenty minutes for a meal break, only one trip home a year, and the exact hours at which they were to return each night: in June and July they were allowed to walk outside in the evenings until ten o'clock; during all other months the curfew was earlier.

Living at the house were Joseph Rowntree's sons, including twenty-one-year-old Joseph and nineteen-year-old Henry Isaac. Joseph was tall and dark with intense features, the natural severity of his own character complemented by years of Quakerly upbringing. His father had taken him to Ireland on a Quaker Relief Mission in 1850 during the potato famine, and the experience had left a lasting impression on him: Joseph remembered how slow starvation turned the young and comely into walking corpses. Numberless unknown dead lay in open trenches or where they had fallen by the side of the road. For the serious-minded Joseph it had been a shocking lesson in the effects of poverty. His younger brother Henry provided a contrast to Joseph's austerity. Somehow the full Puritan weight of Quaker training did not sit quite so readily on him; he had a sense of fun, and could be relied upon to lighten the mood.

While working in Rowntree's grocery shop, George saw at first hand how the family's cocoa business came into existence. Joseph senior had for many years been close friends with a local businessman, Samuel Tuke, who ran a cocoa factory and shop at Castlegate, not far from the Rowntrees' Pavement shop. The business had been in the family for three generations, but when Samuel Tuke became ill in 1857, his sons did not want to take on the cocoa factory. The elder Joseph Rowntree offered to help his friend by placing one of his own sons in the business. As Rowntree's eldest sons were due to take a stake in his grocer's shop, the opportunity to work for the Tukes fell to his third son, Henry Rowntree. In 1860, Henry duly set out to Castlegate to embark on his own career in cocoa.

Not long after this, George Cadbury returned to Birmingham, although he had barely completed three years as an apprentice. Perhaps he was fired up by seeing Henry Rowntree start his new venture, and was eager to begin making his own way. But it also seems likely that John was well aware of his third son's ability and dedication, and needed his help.

To the employees at Bridge Street the two young Cadbury brothers were curiously 'alike and unalike'. Richard was seen as 'bright and happy with a sunny disposition'. He claimed he would be happy simply to rescue the business and turn it around to make a few hundred pounds a year. George was much more driven. In the words of his biographer, Alfred Gardiner, he had 'more of an adventurer's instinct . . . The channel of his mind was narrower and the current swifter.' Despite his ambition, he could see no simple solution to the business's problems. As the brothers deliberated during the spring of 1861 in the gloomy Bridge Street factory, the prospect seemed a dismal one. From their cramped office they could see the empty carts banked up in the yard awaiting orders. It was not immediately obvious what they could do that their father and uncle had not already tried.

The great hope, of course, was to come up with a breakthrough product. They did in fact have something in mind that their father had been working on before family difficulties drained his energy. It was a product very much of the moment, with healthful overtones, called Iceland Moss. The manufacturing process involved blending the fatty chocolate bean with an ingredient that was thought to improve health: lichen. It was then fashioned into a bar of cocoa that could be grated to form a nutritious drink. Richard had a flair for design, and could see the possibilities for launching Iceland Moss. It would be eye-catchingly displayed in bright yellow packaging with black letters that boldly proclaimed the addition of lichen, complete with the image of a reindeer to show how different it was from anything else on the market. He and George hoped to promote the health-giving properties of Iceland Moss, but would the untried combination of fluffy-textured lichen and fatty cocoa bean excite the English palate?

Apart from developing new products, the brothers also had to find new customers. Their father had only one salesman, known at the time as a 'traveller'. His name was Dixon Hadaway, and he covered a vast swathe of the country, from Rugby in the south far up into the Scottish Highlands, visiting grocers' shops to promote the company's variety of cocoa wares. He took advantage of the new railway to cover the long distances between towns, but was often obliged to travel by pony and trap or even on foot. Despite the challenges of getting around, Dixon Hadaway was evidently determined to keep up appearances, and was always smartly attired with a tall top hat and dark tweed coat, although it was invariably crumpled from long hours of travelling. It seems he was appreciated by his customers, who claimed that he was so punctual that they could set their clocks by his visits. But punctuality and enthusiasm alone were not enough to win new orders. People could not be expected to buy Cadbury's goods if they had never heard of them. George was clear: they needed more capital to fund a sales team.

To finance this, the brothers discussed how to manage the business more efficiently. Their solution was to return to their Puritanical roots: 'work, and again work, and always more work'. George enthusiastically planned to cut all indulgence from his life: games, outings, music, every luxury would go. Every penny he earned would be ploughed back into the business. This was harder for Richard, who was planning to marry in July.

A photograph survives of Richard's fiancée, Elizabeth Adlington, whose classic good looks are evident in spite of her serious expression and the limited scope permitted for the enhancement of feminine Quaker beauty. Her face appears unadorned, her hair parted down the middle and pulled back severely. She wears a full skirt and crinoline, covered by a long black cloak and dark bonnet – Quaker fore-fathers deeming this quite sufficient adornment to attract a male. Richard was drawn to her 'bright and vivacious' manner. In preparation for bringing home a wife, he had purchased a house on Wheeley's Road, about two miles from the factory. Spare moments were spent creating a garden, transferring cuttings of his favourite plants from the rockery in his father's garden. 'My little home is beginning to look quite charming now it is nearly completed,' he told his youngest brother Henry. There was just the furniture to buy before the wedding.

During the spring of 1861 the tone of the brothers' discussion changed. As Quakers, they were accustomed to finding answers in silent prayer. They had a duty to their workforce, and there were family obligations to consider. Since their mother had died, their sister Maria had taken her place, caring for the younger members of the family. Now their father was in urgent need of help. They too must listen to the clear voice of conscience, mindful of their debt to man and God. They too must endeavour to do their best. Whatever their misgivings, they had no real choice. In April the two young brothers took over the running of the factory.

There was one last hope. They had each inherited £4,000 from their mother. Determined to save the family dream of a chocolate factory, they staked their inheritance down to the last penny. If they failed to turn the business around before the capital was gone, they would close the factory.

2

Food of the Gods

Richard and George soon found themselves running down their inheritance fast just to keep afloat. The first year was worrying. By the end of 1861 Richard's share of the loss was recorded at £226, and George registered a similar figure. More capital from their inheritance would be needed. Richard, who had the added responsibilities of married life, could not help imagining the next year's losses. Perhaps they were not businessmen. Was this the beginning of a slow and inevitable decline to bankruptcy?

The brothers tried to calculate how long their capital would last. In the absence of any other source of funds, they had to make further cutbacks. Even basic pleasures such as drinking tea and reading the morning paper were now sacrificed. Each day started at six in the morning and did not end until late in the evening, with a supper of bread and butter eaten at the factory. 'This stern martyrdom of the senses,' observed one of George's colleagues years later, 'drove all the energy of his nature into certain swift, deep channels', creating an extraordinary 'concentration of purpose'. Any small diversion or treat was dismissed as a 'snare' that might absorb precious funds.

While George focused on purchasing, policy and development, Richard tackled sales. The infrequent appearances of their traveller, Dixon Hadaway, in the office made a vivid impression on the staff. 'It was a red letter day,' said one office worker. 'It was real fun to listen to his broad Scotch, as we could only understand a sentence here and there.' Hadaway loved his old tweed coat, which he had worn since the Crimean War, 'and I can still remember him extolling the beauties of the cloth and its wearing qualities'.

Dixon Hadaway, the Cadbury brothers' first traveller, whose territory extended from Rugby to northern Scotland.

Richard joined Hadaway on some of his travels, and frequently took out the pony and trap to drum up business. He also hired some additional full-time travellers. Samuel Gordon was to target Liverpool and Manchester, while John Clark, recommended by a Quaker cousin, was hired to take on the whole of England south of Birmingham. Richard sent him first to London, but in a matter of weeks Clark found business there so bad he begged to be transferred back to Birmingham, as he feared he was wasting both his and the firm's time. A letter survives from Richard, urging him not to give up on London and its suburbs:

> We do particularly wish this well worked, as we believe it will ulti-mately repay both us and thyself to do so, and thou may depend if thou dost thoroughly work it, we will see nothing is lost to thee whether with or without success . . . It is important for us both to pull together for we have so much to do to conquer reserve and prejudice, and thou may be assured we will do our part in this in the way of improve-ments in style and quality of our goods.

To cover more ground, George too began to travel, taking on Wales, the Isle of Man and the Channel Islands. Letters from Richard's young wife Elizabeth show that his journeys away from home became more frequent. 'We have come nearly to the end of another day and think of thee as that much nearer returning,' she wrote to him in Glasgow in July 1862, a year after their marriage. 'We shall all be happy together if thou hast had a prosperous time.' In his enthusiasm to increase turnover, Richard himself would go into the warehouse to package the orders, 'not only in the early days when hands were few, but even in his later years'.

During 1862, since both brothers were often away, they hired more office staff. One young worker who showed great promise was William Tallis. Orphaned as a child, he had had very little education, but impressed everyone with his ability and enthusiasm. They also employed their first clerk, George Truman, who recalls 'working, as did Mr George, till eight or nine every night, Saturdays included'. George Truman evidently also tried his hand at selling to the shops in Birmingham. A novice salesman, he generously offered samples for customers to try. These proved so popular that he soon ran out, and returned to the factory 'in great distress' because 'one customer had eaten half his samples!' He was reassured when 'Mr George said he could have as many samples as he wanted and he went out the next day quite happy.'

To address the problem of the product being eaten before it left the factory, a system known as 'pledge money' was put into effect. Each day a penny was awarded to any worker who had managed not to succumb to temptation. Every three months the pledge money was paid out: one particularly abstemious employee remembers he accumulated so much that he was able to buy a pair of boots. Workers were also rewarded for punctuality. For those who arrived promptly at 6 a.m. there was a breakfast of hot coffee or milk, bread and buns.

Unaccountably, the brothers found there was a lack of public enthusiasm for Iceland Moss, in which they had invested their early hopes. They continued to develop new lines of higher quality, introducing a superior Breakfast Cocoa, as shown in their detailed sales brochure of 1862. This was followed a year later by Pearl Cocoa, then by Chocolat du Mexique, a spiced vanilla-flavoured cake chocolate. They improved existing brands such as Queen's Own Chocolate, Crystal Palace Chocolate, Dietetic Cocoa, Trinidad Rock Cocoa and Churchman's Cocoa – a sustaining beverage for invalids. 'So numerous are the sorts,' reported the *Grocer* magazine of these different types of cocoa drink, 'the purchaser is much puzzled in his choice.' So puzzled in fact that no single one of Cadbury's products seemed to excite the palate of the Midlands' growing population.

Richard was keen to find new ways to promote their different types

of nutritious beverage. Apart from notifying the trade through the *Grocer*, in 1862 he designed a stall to exhibit their products at the permanent exhibition in the Crystal Palace at Sydenham in south London. The brothers also paid for a stall exhibiting their wares in the Manchester Corn Exchange. But it was not enough. An elusive 'something' was missing from their products, preventing them from exploiting the clearly delicious cocoa bean. Their travellers returned with disappointing orders, putting the struggling business in further jeopardy.

In battling to save the Bridge Street factory there was one issue that the brothers had failed to tackle. However inventive their new recipes, and however adventurous the palate of the British public, by turning cocoa beans into a drink, they were faithfully following centuries of tradition. Despite its long and colourful history of cultivation, by the mid-nineteenth century the dark cocoa bean was mostly consumed in a liquid form, largely unprocessed and unrefined, as it always had been. The Cadbury brothers, like everyone else, were still thinking along lines rooted in ancient history.

Like many Victorians, Richard Cadbury had a passion for foreign culture and history. With his life circumscribed by long hours of labour in provincial England, he longed to travel beyond Europe. He had been brought up on vivid tales of the exotic lands where cocoa originated, and the history of its cultivation. 'It was one of the dreams of our childhood,' he wrote, 'to sail on the bosom of that mighty river whose watershed drains the greater part of the northern portion of the continent of South America, and to explore the secrets of its thousand tributaries that penetrate forests untrodden by the foot of man.' He was particularly interested in the long and colourful history of cocoa in South America and Mexico, a history that gave intriguing glimpses as to how the bean might best be cultivated and consumed.

Richard had never actually seen a cocoa plantation, and attempted to satisfy his curiosity by collecting stories of explorers. While the traders he had met in Mincing Lane had never been short of anecdotal accounts, he could find out more by corresponding with experts at the tropical botanical gardens in Jamaica, and the Pamplemousse Botanic Gardens in Mauritius. Closer to home, knowledge of tropical species was increasing through the famous glasshouses at the Royal Botanic Gardens at Kew. The magnificent Palm House had recently been completed, and in the early 1860s work was just beginning on the Temperate House. Botanists knew cocoa by its scientific name – *Theobroma cacao*, or 'Food of the Gods' – given to the plant in 1753 by the Swedish naturalist Carl von Linnaeus.

'This inestimable plant,' Richard wrote, 'is evergreen, has drooping bright green leaves . . . and bears flowers and fruit at all seasons of the year.' It flourished only in humid tropical regions close to the equator, and was acutely sensitive even to slight changes in climate. The cocoa pod itself he described as 'something like a vegetable marrow . . . only more elongated and pointed at the end'. In contrast to European fruit trees, the pods grow directly off the trunk and the thickest boughs, from very short stalks, rather than from finer branches. The outer rind of the pod is thick, and when ripened becomes a firm shell. Inside, embedded in a soft, pinkish-white acid pulp, are the seeds or beans – as many as thirty or forty within each pod.

Richard's romantic idea of cocoa plantations was fed by the travel narratives that were occasionally featured in fashionable magazines like the *Belgravia*. One such article described a magical tropical paradise that must have seemed a million miles from Victorian Birmingham. In looking down over the plantation, 'the vista is like a miniature forest hung with thousands of golden lamps . . . anything more lovely cannot be imagined'. Taller trees such as the coral tree, were planted around the cocoa trees to provide shade. In March, the coral trees became covered in crimson flowers, and 'At this season, an extensive plain covered with cocoa plantations is a magnificent object. The tops of the coral tree present the appearance of being clothed in flames.' Passing through the shady walkways of the plantation was like being 'within the spacious aisles of some grand natural temple'.

To harvest the marrow-like cocoa pods, the plantation workers would break them open with a long knife or cutlass. The pale crimson seeds or beans were scooped out with a wooden spoon, the fleshy pulp scraped off, and the beans dried in the sun until they turned a rich almond brown. This method of preparation had remained essentially unchanged for centuries, and is richly interwoven with the history of the Americas.

Richard could not suspect how far into the past this history of cultivation extended. Recent research has revealed that three millennia have elapsed since the Olmec, the oldest civilisation in the Americas, first domesticated the wild cocoa tree. Little survives of the Olmec, who eked out an existence in the humid lowland forests and savannahs of the Mexican Gulf coast around 1500 to 400 BC, save the striking colossal heads they sculpted of their kings. Evidence that these early Mexicans consumed cocoa comes principally from studies in historical linguistics – their word 'ka ka wa' is thought to be the origin of 'cacao'.

When the Maya became the dominant culture of Mexico from around AD 250 they extended the cultivation of cocoa across the plains of Guatemala and beyond. In Mayan culture, the rich enjoyed a foaming, hot, spicy cocoa drink. The poor took their cocoa with maize as a starchy, porridge-like cold soup that provided easily prepared high-energy food. It could be laced with chilli pepper, giving a distinct afterburn, or enhanced with milder flavourings such as vanilla.

Mayan art reveals that cocoa was highly prized. Archaeologists have found images decorating Mayan pottery of a 'Cacao God' seated on his throne adorned with cocoa pods. There is evidence suggesting that Mayan aristocrats were buried with lavish supplies for the afterlife, including ornate painted jars containing cocoa and cocoa flavourings. The earliest image of the preparation of a chocolate drink appears on a Mayan vase from around the eighth century AD, which also depicts a human sacrifice. Two masked figures are beheading their victim, while a woman calmly pours a cocoa drink from one jar to another in order to enhance the much-favoured frothy foam.

'European knowledge of cocoa as an article of diet,' Richard wrote in his survey, 'dates from the discovery of the Western World by Christopher Columbus.' On 15 August 1502, during Columbus's fourth trip to the New World, he reached the island of Guanaja, near the Honduran mainland. His men captured two large canoes and found they were Mayan trading ships, laden with cotton, clothing and maize. According to Columbus's son Ferdinand, there were a great many strange-looking 'almonds' on board. 'They held these almonds at great price,' he observed. 'When any of these almonds fell, they all rushed to pick it up, as if an eye had fallen.' The Europeans could not understand why these little brown pellets should be so valued.

The mystery was solved by the Spanish conquistadors, who arrived in Mexico in 1519. Travelling into the Valley of Mexico, they reached the heart of what was then the Aztec civilisation. It was soon apparent to them that the cocoa bean had special value in Aztec society, since it was used as coinage, and people in the provinces paid tributes to their Emperor, Montezuma, with large baskets of cocoa beans. The Emperor kept a vast store in the royal coffers in the capital city of Tenochtitlan of no fewer than 40,000 such loads: almost a billion cocoa beans. According to one Spanish chronicler, 'a tolerably good slave' was worth around one hundred beans, a rabbit cost ten beans, and a prostitute could be procured for as few as eight.

It is now known that the Aztecs, like the Mayans, used their favourite drink in a number of religious rituals, including human sacrifice. The Aztecs believed that their most powerful gods required appeasement, and prisoners of war had to be sacrificed each day to sustain the universe. In one macabre ritual, the heart of a slave was required to be cut out while he was still alive. The slave was selected for his physical perfection, since until the time of his sacrifice, he represented the Aztec gods on earth and was treated with reverence. According to the Spanish Dominican friar Diego de Duran, who wrote The History of the Indies of New Spain in 1581, as the ritual approached its climax, and the fate of the victim was made known, the slave was required to offer himself for death with heroic courage and joy. Should his bravery falter, he could be 'bewitched' by a special little cocktail to see him through, prepared from chocolate and mixed with the blood of earlier victims and other ingredients that rendered him nearly unconscious.

Some time in the sixteenth century the cocoa bean found its way to Europe, where it was introduced into the Spanish royal household. The Spanish court initially consumed cocoa the South American way, as a drink in a small bowl, and then gradually replaced the maize, or corn, and chillies with sugar, or sometimes vanilla or cinnamon. In time elaborate chocolate pots were developed in which the heavy liquid was skimmed and allowed to settle before pouring, but essentially the Spanish ground the beans in the same way as the inhabitants of the Americas, crushing them between stones, or grinding them with stone and mortar. The result was a coarse powder.

Richard Cadbury found one written account of cocoa preparation in Madrid in 1664 in which one hundred cocoa beans, toasted and ground to a powder, were mixed with a similar weight of sugar, twelve ground vanilla pods, two grains of chilli pepper, aniseed, six white roses, cinnamon, two dozen almonds and hazelnuts and a little achiote powder to lend a red hue. The resulting paste was used to make a cake or block of cocoa which could be ground to form a drink. But whether mixed with maize or corn to absorb the fatty cocoa oils the Mexican way, or blended with sugar, the cocoa oils made the drink heavy and coarse, and cocoa continued to receive a mixed reception in Spain. Josephus Acosta, a Spanish writer at the turn of the seventeenth century, considered the chocolate drink much overrated, 'foolishly and without reason, for it is loathsome . . . having a skum or frothe that is very unpleasant to taste'.

For many years the Spanish dominated cocoa cultivation in the Americas. They introduced it to the West Indies, where on islands such as Trinidad it soon became a staple crop. At first, English pirates raiding Spanish ships did not know what the bean was for. In 1648 the English chronicler Thomas Gage observed in his *New Survey of the West Indies* that when the English or Dutch seized a ship loaded with cocoa beans, 'in anger and wrath we have hurled overboard this good commodity, not regarding the worth of it'.

Gradually, however, during the seventeenth and eighteenth centuries the cocoa bean found its way into the coffee houses of Europe. Its first recorded mention in England appears in an advertisement in the *Public Advertiser* on 22 June 1657: 'At a Frenchman's house in Queensgate Alley is an excellent West India drink called Chocolate to be sold, where you may have it ready at any time at reasonable rates.' Word of the exotic new drink spread across England. In the 1660s, the diarist Samuel Pepys describes enjoying his drink of 'chocolatte' or 'Jocolatte' so much that he was soon 'slabbering for another'. In 1662 Henry Stubbe, royal physician to King Charles II, published A *Discourse Concerning Chocolata*, in which he surveyed the 'nature of the cacao-nut' and extolled the health benefits of the drink. Taken with spices it could relieve coughs and colds, and strengthen the heart and stomach. For anyone 'tyred through business' Stubbe heartily recommended chocolate twice a day. It could even serve as an aphrodisiac. White's famous Chocolate House opened in 1693 in St James's, London.

Cocoa continued to gain in popularity, principally as a drink prepared in the Mexican way, but it was also added as a flavouring to meat dishes, soups and puddings. In Italy a recipe for chocolate sorbet survives from 1794 – once the mixture was prepared, 'the vase is buried in snow layered with salt and frozen'. For the more adventurous palate, Italian recipes from the period listed cocoa as an ingredient in lasagne, or even added it to fried liver. While most European preparations were 'rough . . . and produced poor results', according to Richard Cadbury, 'France developed a better system for roasting and grinding.' The French *confiseurs* got straight on with the sweet course; no messing with chillies, curries or fried liver. By the nineteenth century, French confectioners were winning a reputation for their exquisite hand-crafted sweets made from chocolate: delicious mousses, chocolate cakes, crèmes and dragées and chocolate-coated nuts.

With the wheels of European commerce and consumerism driving demand, in the Americas the cultivation of the cocoa bean was gradually extended beyond Mexico and Guatemala, reaching south to the lower slopes of the Andes in Ecuador, the rolling plains of Venezuela and into the fringes of the Amazon rainforest of Brazil. In the Caribbean, cocoa plantations were established on Jamaica, St Lucia and Grenada as well as Trinidad, which became a British colony in 1802.

By the mid-nineteenth century, despite the growing interest in the cocoa bean in Europe, it remained expensive, and principally a novelty for the wealthy. 'When we take into account the indifferent means of preparation,' concluded Richard Cadbury, 'we can hardly be surprised that it did not come into general favour with the public.' As Richard and George struggled to save their business, the full potential of *Theobroma cacao* had yet to be revealed. The unprepossessing little bean offered only a tantalising promise of prosperity.

⋖⟡⋗

In the early 1860s George and Richard hardly needed to undertake the charade of stocktaking, which they did twice a year. They knew their business did not thrive. The Cadbury brothers' wide variety of different cocoas did not excite the nation's tastebuds. 'We determined that we would close the business when we were unable to pay 20 shillings in the pound,' George said, determined to honour all their financial agreements in full. He admitted the stocktaking was 'depressing', but nonetheless he and Richard thrived on the challenge: 'We went back again to our work with renewed vigour and were probably happier than most successful men.'

The struggle brought the brothers closer together. Quite apart from sharing the responsibility and the burden with George, Richard proved to be a delight as a partner. He was 'very good natured and constantly up to practical jokes and fun of various kinds', George wrote, 'so that one almost doubts whether immediate success in a business is a blessing'. Workers too recalled Mr George and Mr Richard with a 'cheery smile', although they knew 'the Firm was in low water and losing money', and 'at one stage expected any day to hear that the works were to be shut'.

Despite his outward calm, as losses mounted Richard privately formed a list of everything he owned, noting the price each item would fetch if it had to be sold at auction. The birth on 27 September 1862 of his first son, Barrow – named after Richard's mother's family – was great cause for celebration, but Richard knew the financial security of his young family was uncertain, and the brothers fully intended to shut the business rather than risk defaulting on any money owed and accruing debt. The stocktaking at the end of the second year was particularly gloomy. By Christmas 1862, the Cadbury brothers' losses had escalated to a further £304 each.

But for Richard and George there was another motive that went well beyond personal gain. Business was not an end in itself; it was a means to an end. As Quakers, they had a far greater goal to fulfil.

3

The Root of All Evil

Richard and George Cadbury shared a vision of social justice and reform: a new world, in which the poor and needy would be lifted from the 'ruin of deprivation'. For generations, the Cadburys had been members of the Society of Friends, or Quakers, a spiritual movement started by George Fox in the seventeenth century. In a curious irony, the very religion that inspired Quakers to act charitably towards the poor also produced a set of codes and practices that enabled a few thousand close-knit families like the Cadburys to generate astounding material rewards at the start of the industrial age.

Richard and George had been brought up on stories of George Fox, and many of the values, aspirations and disciplines that shaped their lives stemmed from his teachings. Born in 1624, the son of a weaver from Drayton-in-the-Clay (now known as Fenny Drayton) in Leicestershire, Fox grew up with a passionate interest in religion at a time when the country had seen years of religious turmoil; he went to 'many a priest looking for comfort, but found no comfort from them'. He was appalled at the inhumanity carried out in the name of religion: people imprisoned or even executed for their faith. Disregarding the danger following the outbreak of the Civil War in 1642, he left home the following year in some torment, and set out on foot for London. At the age of just nineteen, Fox embarked on his own personal quest for greater understanding.

During these years of travel, 'when my hopes . . . in all men were gone', he had an epiphany. The key to religion was not to be found in the sermons of preachers, but in an individual's inner experience. Inspired, he began to preach, urging people simply to listen to their

George Fox, the founder of the Quaker movement.

own consciences. Because 'God dwelleth in the hearts of obedient people,' he reasoned, it followed that an individual could find 'the spirit of Christ within' to guide them, instead of taking orders from others.

Fox blazed a trail across England, preaching against the rituals and outward forms of religion, even the standard forms of prayer and the sacraments. All these he regarded as trivial accessories; irrelevant, possibly even hindering a union with God. These outward symbols of truth, he reasoned, obscured or distorted the real truth, which could be found within each one of us. He spoke out against the corruption within the Church, arguing for social justice and a more honest and immediate form of Christianity. These views put him in direct opposition to the political and religious authorities. If an individual was listening to the voice of God within himself, it followed that priests and religious leaders were needless intermediaries.

Fox was perceived as dangerous, and his preaching blasphemous to established Churches. Even the similarly-minded Puritans objected to him. They too adhered to a rigorous moral code, high standards of self-discipline and a disdain for worldly pursuits, but Fox's emphasis on the direct relationship between a believer and God went far beyond what most Puritans deemed acceptable. In emphasising the importance of an individual's experience, Fox appeared contemptuous of the authorities, and mocked their petty regulations. For example, he would not swear on oath. If there was only one absolute truth, he reasoned, what was the point of a double standard, differentiating between 'truth' and 'truth on oath'?

The authorities were exasperated that he declined to pay even lip-service to the class structure, and went so far as to claim that all men and women are equal. To give tangible form to his thoroughly modern message, Fox addressed everyone as 'thou', not the more respectful 'you' that others used to acknowledge the upper classes. He rejected any outward signs of status or authority. Regardless of wealth, a person should dress simply, with restraint and without extravagance. As for the doffing of hats to indicate respect for those of higher rank, in his *Journal* he made his position completely clear: a Quaker kept his hat on.

In 1649 Fox crossed one magistrate too many, and was thrown into jail in Nottingham, 'a pitiful stinking place, where the wind brought in all the stench of the house'. The following year he was jailed for blasphemy in Derby, where a Justice is believed to have been the first to use the term 'Quaker', to mock Fox and his followers. He scoffed at the idea expressed in their meetings that they should remain silent until moved to speak, 'trembling at the word of God'. Despite its origins as a term of abuse, the name 'Quaker' soon became widespread.

Fox was imprisoned several times, but the Quaker movement continued to gain momentum. It is estimated that during the reign of Charles II, 198 Quakers were transported overseas as slaves, 338 died from injuries received defending their faith, and 13,562 were imprisoned. Among them were some of Richard and George's fore-bears, including one Richard Tapper Cadbury, a wool comber who was held in Southgate prison in Exeter in 1683 and again in February 1684.

By the end of Fox's life in 1691 there were 100,000 Quakers, and the movement had spread to America, parts of Europe and even the West Indies. Fox had established a system of regular meetings for Friends to discuss issues and formalise business: the regional Monthly Meeting, the county Quarterly Meeting and a national Yearly Meeting. Key decisions made at these meetings were written down, and these records became known as the *Advices*. By 1738 they had been collated by clerks, transcribed in elegant longhand, and bound in a green manuscript volume, *Christian and Brotherly Advices*, which was made available to Friends' Meetings across the country. This set out codes of personal conduct for Friends, under such headings as 'Love', 'Covetousness' and 'Discipline'. A section on 'Plainness', for example, encouraged Quakers to cultivate 'plainness of speech, behaviour and apparel'. A Friend's clothing should be dark and unadorned; even collars should be removed from jackets, as they were deemed too decorative.

The strict rules of the Quakers dictated that anyone who married outside the society had to leave. Consequently, Quaker families tended to intermarry, resulting in a close-knit community across Britain of several thousand families. Generations of Quakers had come through years of persecution and suffering with a sense of solidarity, and these bonds were also forged by friendship, marriage, apprenticeships and business. As the Industrial Revolution was gathering speed, this solidarity and self-reliance generated a new spirit of enterprise. At a time when there was no such thing as a national newspaper, the Quakers meeting regularly in different regions across Britain enjoyed a unique forum in which to exchange ideas.

In 1709 Abraham Darby, a Quaker from Shropshire, pioneered a method of smelting high-grade iron using coke rather than charcoal. His son, Abraham Darby II, improved the process, replacing the traditional horse pumps with steam engines to recycle water, and refining techniques for making quality wrought iron. The Darbys manufactured the world's first iron bridge, iron railway tracks and wheels at their foundry at Coalbrookdale in Shropshire. Their roaring furnaces drew visitors from miles around to observe the striking spectacle of flame, smoke and machine. The younger Darby's daughter wrote in 1753 that the noise of 'the stupendous bellows' was 'awful to hear'.

Such advances fuelled the development of the iron industry, which drove the wheels of the Industrial Revolution. In Sheffield, the Quaker inventor Benjamin Huntsman developed a purer and stronger form of cast steel. The Lloyds, a Welsh Quaker family, moved to Birmingham to create a factory for making iron rods and nails. In Bristol, a Quaker cooperative launched the Bristol Brass Foundry. By the early eighteenth century Quakers ran approximately two-thirds of all British ironworks.

Railways accelerated the pace of change, and a Quaker was responsible for the world's first passenger train. In 1814 a meeting with the engineer George Stephenson inspired Edward Pease, a Quaker businessman, to build the Stockton and Darlington Railway, and on 27 September 1825 the first steam-hauled passenger train travelled twelve miles to Stockton on what became known as the 'Quaker Line'. Numerous Quakers were involved in financing and directing railway companies. Even the railway ticket and stamping machine was devised by a Quaker, Thomas Edmonson, as was the timetable itself, *Bradshaw's Railway Times*, devised by George Bradshaw.

There seemed no limit to the number of new ideas from Quaker businessmen. Chinaware, originally imported by the East India Company, sparked developments in pottery and porcelain. In Plymouth, William Cookworthy introduced a new way to make fine china using Cornish china clay. In Staffordshire, Josiah Wedgwood launched his pottery business. Enduring shoe businesses were founded by Quakers: K shoes in Kendall by John Somervell, and James Clark in Street in Somerset established the firm that still bears his name. The Reckitts started their business in household goods, while the Crosfields were soap and chemical manufacturers whose company evolved into Lever Brothers. The roll call of Quaker entrepreneurs resounds through the centuries, with names like Bryant and May, who designed a safer form of matches; Huntley and Palmer, who started a biscuit business in Reading; and Allen and Hanbury, who developed pharmaceuticals.

Banking too was built on Quaker virtue. At a time of little financial regulation, according to the writer Daniel Defoe the activities of many eighteenth-century financiers were 'founded in Fraud, born of Deceit, nourished by Trick, Cheat, Wheedle, Forgeries, Falsehood' (not totally dissimilar to some twenty-first-century banks, some might argue). The Quaker traders stood out as being quite different. Customers learned to rely on typical Quaker attributes: skilled bookkeeping, integrity and honesty served up by sober Bible-reading men in plain dark clothes. In the seventeenth and eighteenth centuries, local Quaker businesses began providing a counter in their offices that offered banking services. By the early nineteenth century this practice had blossomed into seventy-four Quaker banks, one for almost every large city in Britain: James Barclay formed Barclays Bank in London, Henry Gurney established Gurney's Bank in Norwich, Edward Pease formed the Pease Bank in Darlington, Lloyd's Bank was started in Birmingham, Backhouse's Bank grew across the north, Birkbeck flourished in Yorkshire, the Foxes set up in Falmouth, the Sparkes in Exeter, and many more. By the time Richard and George Cadbury were born, Quaker banks, founded on a unique and trusted set of values, formed a solid network across the country.

Underpinning all this, the core Quaker beliefs and traditions, and the independence of spirit that went with them, flourished. As its members' banks and businesses grew, the Society of Friends continued to exchange views in meetings across the country. The stoic independence, self-discipline and questioning rebelliousness fashioned over a century was now channelled into the spirit of enterprise that fuelled the furnaces and mills of the Industrial Revolution.

But there was something else unique that guided Quakers in business from the earliest days of the movement. The original *Christian and Brotherly Advices* of 1738 included a section on 'Trading'. This highlighted situations that a Friend might encounter in business, and how to deal with them. It marks the foundation of business ethics built on truth, honesty and justice: values that would form the basis of Quaker capitalism.

Central to the advice was that a Quaker must always honour his word:

- *That none launch forth into trading and worldly business beyond what they can manage honourably and with reputation among the Sons of*

Men, so that they may keep their word with all Men; that their yea
may prove their yea indeed, and their nay, may be nay indeed; for
whatever is otherwise cometh of the Evil One . . . and brings Dishonour
to the Truth of God.

Quakers entering into business were encouraged to keep written
accounts, since accurate and thorough bookkeeping helped avoid
errors of judgement.

• *It is advised that all Friends that are entering into Trade and have not*
  stock sufficient of their own to answer the Trade they aim at be very
  cautious of running themselves into Debt without advising with some of
  their Ancient and Experienced Friends among whom they live.

Above all, Quaker elders, many of whom were in trade themselves,
were keen to prevent any 'Great Reproach and Scandal' that might
damage the reputation of the Society:

• *It is advised . . . that all Friends concerned be very careful not to contract*
  Extravagant Debts to the endangering and wronging of others and their
  families, which some have done to the Grieving Hearts of the upright,
  nor to break promises, contracts and agreements in the Buying and
  Selling or in any other lawful Affairs, to the injuring themselves
  and others, occasioning Strife and Contention and Reproach to Truth and
  Friends.

The local Monthly Meetings across the country were not only a
forum for exchanging ideas: Quakers were urged to 'have a Watchful
Eye over all their Members'. If they found anyone 'Deficient in
Discharging their Contracts and just Debts', they were charged with
'launching an Inspection into their Circumstances'. Should the transgressor fail to heed honest advice, 'Friends justifiably may and ought
to testify against such offenders.' Accordingly, Friends collaborated in
their local communities to help one another achieve high standards
of integrity in trade.

For those who failed to comply, there were further words of guidance. Despite the *Advices'* exhortation to Friends to aspire to
'Truthfulness and Perseverance in Godliness and Honesty', 'to our

Great Grief we find there are fresh instances of Great Shortness in coming up in the Practice Thereof, particularly by some injuriously defrauding their Creditors of their Debts'. This had led to 'Grievous Complaints'. The *Advices* urged further discipline to deal with those 'Evil Persons' who proved 'base and unworthy'. Firstly, the rule which prohibited Friends from suing one another could be waived. Secondly, Monthly Meetings had the power to investigate cases and 'speedily set righteous Judgment upon the head of the Transgressor'.

Discipline could be severe for any members who were unable to meet the ethical standards required, or who acted imprudently in business. With the astonishing success of Quaker businesses and banks during the Industrial Revolution, protecting the good name of the Society became more important. Those who repeatedly failed to demonstrate the high ethical conduct required of a Quaker tradesman could be 'disowned' by the Society. This was seen as a harsh punishment, with the offender excluded from the local Quaker community and recognised publicly as a thief or a cheat.

Effectively as early as 1738, Quakers had a set of specific guidelines for business, which endeavoured to apply the teachings of Christ to the workplace. Straight dealing, fair play and honesty would form the basis of Quaker capitalism, and for those who fell short, there were rules of discipline. These guidelines were supplied to clerks at the Monthly and Quarterly Meetings, and were refined and formally updated every generation. They provide a snapshot of changing ethical concerns as the Industrial Revolution gathered momentum. For example, when the trading guidelines were updated in 1783 in the *Book of Extracts*, Friends were warned against a 'most pernicious practice' which could lead to 'utter ruin': the use of paper credit. This was considered 'highly unbecoming', falling far short of 'that uprightness that ought to appear in every member of our religious society'. The 1783 *Extracts* warned unequivocally that this practice was 'absolutely inconsistent with the truth'.

The 1738 *Advices* and 1783 *Extracts* were updated once again in 1833 into the more formal *Rules of Discipline*. By this time, material prosperity presented another issue to exercise the minds of Quaker elders. Was it right for a religious person cultivating plainness and simplicity to accumulate wealth? 'We do not condemn industry, which we believe to be not only praiseworthy but indispensable,' noted the

Rules of Discipline, but 'the love of money is said in Scripture to be "the root of all Evil"'. The guidelines urged 'Dear Friends who are favoured with outward prosperity, when riches increase not to set your hearts upon them.' The work ethic was entirely acceptable, but accumulating riches for oneself was not.

In 1861, as Richard and George Cadbury embarked on their business life, Quaker guidelines were updated once again, in *Doctrine, Practice and Discipline*. By now the section on trade had become a sophisticated set of rules, under the heading 'Advice in Relation to the Affairs of Life'. These covered a wide range of issues: honesty and truthfulness, plain dealing, fair trading, debt, seeking advice from fellow Friends, inappropriate speculation, discipline, and much more. With an increased number of Quakers experiencing worldly success, there was even a section for the children of rich Quakers, to ensure that they were not corrupted but fixed 'their hopes of happiness on that which is substantial and eternal'. The love of money was 'a snare, which is apt to increase imperceptibly . . . and gradually withdraw the heart from God'.

Richard and George Cadbury's entire worldview was shaped by Quaker values. They moulded their early childhood experiences, their learning as apprentices, their social and marriage opportunities, their choice of career, and their all-encompassing view of the wider purpose of their chocolate business.

From the earliest years they had seen their father endeavour to apply Quaker ideals in the community. According to George's biographer Alfred Gardiner, John Cadbury was deeply concerned about society's 'savage indifference to the child'. This was before Charles Dickens made the Victorian public finally take notice of the plight of the 'Parish Boy' and the 'little workus', in his description of child criminals in *Oliver Twist* in 1837. In the 1820s, when John was developing his shop on Bull Street, it was not uncommon for children to be carted off from the workhouses 'like slaves to the cotton mills of Lancashire or to the mines', to be used there as if they were disposable. John was horrified by the misery and degradation of children trapped in a life of slavery.

His greatest outrage was reserved for the 'barbarous practice' of using workhouse boys as young as five as chimney sweeps. Some chimneys were as narrow as seven inches square, and the children could only be induced to climb up inside by straw being lit beneath them, or being prodded with 'pins'. Before they grew too big to be useful, many suffered twisted spines or damaged joints, or were maimed by falls or burns. When John was informed of a machine that could clean chimneys, he 'had the courage to call a meeting of Master Sweeps in the Town Hall', reported the *Daily Gazette*. But his demonstration of the new machine met with strong opposition, as the sweeps were convinced they got better results using boys. After years of campaigning he was delighted when legislation was eventually introduced banning the use of climbing boys.

George and Richard also saw their parents become passionately involved in another major social issue of the time: alcoholism. The consumption of gin had become widespread in the eighteenth century, when many traditional pubs and alehouses were replaced by gin shops which promised oblivion with the tantalising slogan 'Drunk for a penny. Dead drunk for two pence. Clean straw for nothing.' This 'liquid fire', in the words of London magistrate John Fielding, led to nothing less than 'hell'. The painter and social critic William Hogarth wrote of gin causing 'Distress even to madness and death'. Reports of children dying of neglect from their drunken parents were commonplace. There were even accounts of children being killed by their parents, their clothes sold for a pittance for more gin.

By Victorian times the gin shops had become 'gin palaces', whose gilded interiors and warm gaslights seduced workers pouring out of mills and mines on payday. As a member of the Board of Street Commissioners in Birmingham, John Cadbury saw at close hand the squalid reality for those beguiled by such temptations, and he and Candia became keen supporters of the Temperance Movement. In 1834, John publicly signed up to become a total abstainer, and he and Candia vigorously took on the town's drinkers, including even the 'Moderation Society', which tolerated modest drinking, with a 'Total Abstinence Plan'. His researches showed that one house in every thirty in Birmingham 'was dedicated to the sale of intoxicating drinks', and that of the city's 6,593 drunkards, one in ten died each year.

John felt a strong desire to help his fellow citizens who preferred the oblivion of gin to any further struggle with life. In meetings across the town he told them that the money they saved by giving up alcohol could buy a better diet, and compared the hearty meals of roast meat and a quarter loaf that an abstainer could afford to those of a drinker on the same wage, who could bring home little more than a penny loaf. To convince his audience how little goodness there was in a gallon of ale, he lit a saucer of alcohol and watched it vanish in flames. As for the barley in a gallon of beer, he told them, it could be used to make something much more nutritious. At this point he would pass round some of Candia's barley puddings to demonstrate the joys of repudiating drink.

Candia too became personally involved, and for many years 'did go from house to house and court to court, circularising tracts and conversing with the people to induce them to discontinue the drinking usage and practices'. John later wrote that she saw it as her 'duty to seek a personal interview with the landlords of public houses, spirit and beer shops'. These visits were not always appreciated: sometimes 'she was met by rude and coarse remarks'. But often 'much tenderness of feeling was displayed, tears flowed freely, with the expression of the desire to get out of the trade'. She almost certainly picked up the consumption that killed her from these trips, but even as her health

John and Candia Cadbury and their family in 1847.

was declining she continued her crusade, and insisted on making more than two hundred visits to publicans. Richard and George remembered her concerned interest in the children of the poor, who suffered from the consequences of having drunken parents. The Cadburys' Total Abstinence Plan was so successful that, according to John, 'very soon the "Moderation Society" sank into oblivion'.

One tale recorded in Richard Cadbury's *Family Book* concerns the old Birmingham workhouse, which was then at the corner of Lichfield Street and Steel House Lane. When John arrived there for his first meeting as an Overseer of the Poor, he was dismayed to find that the distinguished committee, in true Dickensian style, met once a month for a 'sumptuous repast', with members filling themselves with 'the choicest delicacies', washed down with brandy, before 'attending to the shivering paupers outside'.

Bubbling over with righteous anger, John set out to expose the 'illegality and iniquity' of such banquets. Evidently this was met with some disfavour. In the heated debate that followed, one old gentleman who had never been known to speak on any former occasion was stirred to rise to make a brief but pithy point: 'I spakes for the dinners!' Needless to say, John managed to get the practice stopped.

He also served on the wonderfully-named Steam Engine Committee, which was responsible for tackling what he saw as the 'serious evil' of smog and smoke. As chairman in the 1840s, he gathered data on the Birmingham chimneys that were emitting the greatest volume of

The Birmingham workhouse.

dense black smoke, and put pressure on their proprietors to take action. He was also chair of the Markets and Fairs Committee, which dealt with matters such as unwholesome meats and fraudulent trading. And he won funds as governor of the Birmingham General Hospital to develop its facilities. There was, according to the *Daily Gazette*, a widespread belief that the poor were operated on to advance medical knowledge, and John would periodically attend surgeries 'to prevent any unnecessary cruelty to patients of the poorest class'.

Their parents' example of patient and helpful concern for society's less privileged members was a mantle that George and Richard accepted as an absolutely normal Quaker duty. Not only did they see it as their moral responsibility to improve the plight of those living in the industrial slums, but saving the chocolate factory also held out the promise of providing employment, thus helping the entire community. Even more fundamental, by developing and promoting cocoa as a drink that everyone could afford, they aimed to provide a nutritious alternative to alcohol.

Despite their diminishing inheritance, George and Richard persevered with their efforts to keep the company afloat. George saw the relationship with the employees as key. Sitting in the stock room at 6 a.m. over breakfast, he encouraged workers to discuss issues in their lives, and tried to help with their education, reading aloud to them and exchanging views on topics of interest or stories from the Bible. By today's standards such actions might seem paternalistic and even intrusive, but at a time when many people could not read, they were greatly valued. Many staff members spoke of their enjoyment of these small meetings, which were 'more like family gatherings'. One youth named Edward Thackray recalled how honoured he felt when Mr George called him into his office 'and they knelt together in prayer over some weighty business question'.

The brothers' interest in the workers was also practical. In spite of their losses, George and Richard pressed ahead with plans to increase wages, with a new payment structure that tripled women's pay. A staff fire brigade was organised, which fortunately was never tested by a serious fire in the chocolate works. The brothers introduced the novel idea of a 'Sick Club' to help pay the wages of staff who had to take leave for illness. There was an evening sewing class once a week at the factory, during which George read to the group. The firm's 'bone-shaker' bicycle – with iron-rimmed wheels and no springs – was extremely popular, and any employee could take it home if they had learned to ride it. Richard and George were among the first employers in Birmingham to introduce half days on Saturdays and bank holidays.

They even took the staff on leisure outings. According to the *Daily Post* of 21 June 1864, 'On Thursday last, Messrs Cadbury brothers . . . with commendable liberality took the whole of their male employees on a delightful trip to Sutton Park. The afternoon was spent by some in playing cricket . . . and others rambling through the park enjoying the invigorating air.' At five o'clock the whole company 'sat down to a substantial tea which was duly appreciated'. There was cricket in the summer, and during the winter, 'when work was a bit slack', reported office worker George Brice, 'the appearance of Mr George with his skates was a sure sign that we were to be the recipients of his favour in the shape of a half day's skating'.

As his business experience grew, George was conscious that a paternal responsibility for the firm's employees was falling gently on his shoulders, quite naturally from friendly daily contact. The welfare of the staff was woven into the brothers' lives. The factory was not just a business, it was a world in miniature, and an opportunity to improve society. In the middle of the great big sinful city, George would create a perfect little world, a 'model chocolate factory'.

But first he had to make a profit.

4

They Did Not Show Us Any Mercy

No amount of prayers or hymns could solve one problem: the Cadbury brothers faced stiff competition. The other English cocoa manufacturers 'showed no mercy', claimed George, although they spoke with a friendly face and a reasonable voice. From the cramped offices of the Cadburys' modest factory, their rivals looked unassailable. They sailed expertly on the great seas of commerce, making it look easy, inviting, like an adventure.

In London, the Taylor brothers claimed to be one of the largest cocoa and mustard manufacturers in Britain. They were making very similar products to the Cadbury brothers from their huge cocoa, chicory and mustard manufactory on a large site between Brick Lane and Wentworth Street in Spitalfields. A picture of their works proudly depicted on their sales brochure showed a vast complex of factory buildings with smoking chimneys, and horses and carriages gaily travelling to and fro. Their sales list boasted more than fifty different types of drinking cocoas, including all the familiar lines of Victorian England. Established in 1817, they had gained considerable expertise in cocoa preparation, and claimed their technical knowhow guaranteed the removal of any noxious, greasy oiliness from their delicious products. The firm was huge and, surrounded by the ever-growing London population, continued to grow seemingly unstoppably.

The Taylors were not the Cadbury brothers' only competition in the capital. There was also Messrs Dunn and Hewett of Pentonville, who sold an enterprising range of cocoas that included Vanilla Shilling Chocolate sold 'unwrapped', various types of Chocolate Sticks in tin foil, and a curious Patent Lentilised Chocolate sold in 'half pound canisters'. The early chocolate drinks made with powdered lentils, tapioca, dried peas or sago to mop up the cocoa fats were possibly not for connoisseurs, but these thick, rich cocoa-soups did satisfy the untried tastebuds of many a Londoner. And for the really hard-up, Dunn and Hewett promoted a slightly fatty 'Plain Chocolate Sold in Drab Paper'.

In addition to London there were regional centres of chocolate production, notably at York, where the apprentice George Cadbury had himself witnessed the daring and confidence of Henry Rowntree on his entry into the world of the chocolatier. Within two years of starting at Tuke & Co., Henry was in a position to buy out the company's entire cocoa division.

Henry could see that the Tuke premises, situated in the narrow, winding Castlegate in the heart of the old city of York, were too cramped for the expansion he planned. In buccaneer spirit he bought for £1,000 what he called a 'wonderful new machine' for grinding beans. Included in the sale was a motley collection of collapsing buildings at Tanners Moat near the centre of York, which he optimistically described as his chocolate factory: an ancient ironworks, an alehouse, and several cottages in various stages of disrepair, all of them practically falling into the putrid-smelling River Ouse. Henry

The Rowntree factory at Tanners Moat in York in 1901.

explored the ruinous site with enthusiasm, smelling only chocolate as he glimpsed the river's black and treacherous water.

The company's leading brand had been Tuke's Superior Rock Cocoa, which Henry duly relabelled as Rowntree's Prize Medal Rock Cocoa after it won a prize at a local fair. To promote his wares, Henry extolled the virtues of his Rowntree's Rock Cocoa compared to rival brands. He evidently had a sense of humour and his use of a quotation from Deuteronomy was no doubt appreciated by those listeners who knew their Bible: 'For their Rock is not as our Rock, even our enemies themselves being judges.' As Henry embarked on transforming the Tanners Moat works into a modern factory that could churn out his Rock Cocoa to sell across England, George Cadbury knew that this rival had the determination to succeed.

There were other firms in York poised to benefit from the arrival of the railway. By the middle of the nineteenth century, the twelve trains a day leaving London were delivering 275,000 visitors to York in a year. Joseph Terry and his brothers, who had inherited their father's confectionery business in 1854, took advantage of this new opportunity. Starting back in 1767, their forebears had sold boiled sweets and candied peel to the rich from their enticing sweet shop not far from Rowntree's grocer's shop. Opening the shop door ushered the customer into a magical Hansel and Gretel world of sugared strawberries, raspberries, lemons and oranges.

With the arrival of the railways, Joseph Terry was soon selling to customers in more than seventy-five towns across the Midlands and the north of England. To meet the rising demand, he moved his manufacturing in 1862 to a larger site beside the River Ouse, just outside York's city walls, to which the twice-weekly steam packet brought deliveries of exotic fruits and cocoa. In the 1860s Terry was looking closely at how to diversify his range by making more use of cocoa in chocolate-covered nuts and sweets.

But the Cadburys – and the other cocoa manufacturers – faced their stiffest competition from a giant Quaker concern in Bristol: Fry and Sons. The Frys ran the largest cocoa works in the world, so large it was fast shaping the city of Bristol. Their factory was the size of a small town, their sprawling works easily accommodating all the varied processes of production. This was cocoa-making for England. The Cadburys' little plant could not compete.

Rather than being daunted by the size of the Fry enterprise, George Cadbury was intrigued. 'I never looked at the small people or the people who had failed,' he declared. 'I wanted to know how men succeeded, and it was their methods I examined, and if I thought them good, applied.' Through the Quaker network he was able to approach the Frys in Bristol, and found one partner, Francis James Fry, who was prepared to take him under his wing. Francis Fry and George Cadbury formed a loose alliance of English cocoa-makers which met, for convenience, in the London offices of the Taylor brothers.

'I suppose we had some energy,' George recalled years later, 'for Francis James Fry elected to go round with me to see the Cocoa and Chocolate Manufacturers.' George was surprised by this, remarking, 'I was a young man in a small business compared to his.' A year older than George, Francis James was the fourth generation of his family to run the firm. With Fry's sales approaching a colossal £100,000 a year, he must have felt secure in the knowledge that the young Cadbury brothers were no threat.

George Cadbury had everything to learn about the development of a family firm from his Fry counterpart – and he did. The Bristol firm, he noted, from its earliest years, had had an outstanding reputation for innovation.

⊷◦§✿❧◦⊶

The story of the house of Fry opens in Bristol at a time when the city had more in common with the Tudor period than the modern world. Born in 1728 into a Wiltshire Quaker family, Francis James Fry's great-grandfather, Joseph Fry, who had trained as an apothecary, came to Bristol as a young man seeking an opportunity.

At the time, Bristol was the West Country hub for trade, and as a port was second only to London. On the quayside the harbour opened onto a forest of rigging and sails from a multitude of ships arriving from and departing for the New World. The port was packed with sailors, slaves and merchants, the air heavy with the scent of rum and tar, and marvels from the New World such as sugar and cocoa that were unloaded into wagons and warehouses. In the eighteenth century the Flying Coach, drawn by relays of six to eight horses, made it possible to reach London in two days.

Joseph Fry, a sober figure in his plain Quaker clothes, took a tiny shop in Small Street and began his apothecary business in 1753, at a time when it was still customary for such businesses to keep jars of leeches in the window. As a sideline to his pills and potions he sold cocoa, which he promoted as a health drink, and a highly nutritious alternative to alcohol. Fry's chocolate drink became popular in fashionable nearby Bath, where smart coffee houses were soon selling it to the aristocracy.

In just eight years, Joseph Fry was in a position to take over the leading cocoa manufacturer in the area, Walter Churchman. While Fry's cocoa drink consisted of the oily cocoa flakes and powder in suspension in liquid, Churchman's was clearly superior. Its secret rested on a patent he had taken out in 1729 for 'an invention and new method for the better making of chocolate by an engine'. This was a water-powered machine that enabled him to create a much finer cocoa powder than anyone else. Once Fry had secured the recipe, his Churchman's Chocolate became very popular.

Joseph Fry was inventive, and seized his chances to develop his business. By 1764 he had agents promoting his products in no fewer than fifty-three towns, and was in a position to open a warehouse in London. In 1777 he moved his cocoa manufactory to larger premises in the fashionable Union Street, then on the banks of the River Frome, and used water power to drive the cocoa-grinding mills. His business interests were many and varied, and under his concerned gaze and industrial 'green fingers' everything he touched flourished. He owned a share in the Bristol China Works, created a type foundry in London, was a partner in a large soap- and candle-making business in Bristol, and bought a share of a chemical works in Battersea. This was some feat for a businessman before the age of railways, telegraphs and telephones, and with little means of communication beyond the Flying Coach and the Penny Post.

In 1795 Joseph's son, Joseph Storrs Fry, inherited the cocoa business and continued to develop the Union Street works. Since the water flow from the River Frome was not reliable, he took the remarkable step of installing one of James Watt's first steam engines. To the astonishment of the workers, this clanking, hissing mechanical marvel transformed cocoa production, and was soon regarded as 'one of the wonders of the World'. According to Fry's records, the steam power from this engine was diverted 'by means of a vertical shaft carried up through the factory' to the third floor, where it turned Britain's first 'mechanically driven machine for grinding Cocoa Nibs'. News of the novel idea of using a Watt steam engine for food manufacture prompted comments from across the country. 'We are credibly informed', marvelled the *Bury and Norwich Post* on 6 June 1798, that 'Mr Fry of Bristol has one of these Engines – improved by an ingenious Millwright of the city – for the *sole purpose* of manufacturing Cocoa. It is astonishing to what variety of manufactures this useful machine has been applied!'

Apart from installing a steam engine to grind the cocoa beans,

Fry's grinding machines.

Joseph Storrs Fry received a patent from George III to build a new kind of machine to roast them, which he installed in the factory next door. Doubtless he was gratified to find *The Times* full of praise on 8 August 1801 for the 'excellent articles produced from his celebrated manufactory'. By the time George Cadbury's father was opening his tea and chocolate shop in Birmingham in 1824, the Frys were using nearly 40 per cent of the cocoa imported into Britain, and their annual sales had risen to £12,000.

In 1835 the business passed to the third generation of Frys. Brothers Joseph II, Francis and Richard continued to develop the site on Union Street, and pioneered new brands. They launched Pearl Cocoa, which countered the heavy oiliness of the cocoa drinks of the time with the addition of arrowroot, which absorbed the oils. Since Pearl Cocoa also contained less costly ingredients like molasses and sugar, it could be cheaply priced to attract poorer households, and became a huge seller. Homeopathic Cocoa took advantage of the burgeoning interest in health. For the upmarket consumer the Fry brothers introduced a finely ground Soluble Cocoa which was slightly less gritty. All these products could be sold for a fraction of what it cost to manufacture them a hundred years earlier, when a pound of their grandfather's best cocoas cost over seven shillings, almost as much as the average farm worker received for his weekly wage. These new variations cost around one shilling per pound, while the Frys' workers were paid ten shillings per week.

The Fry family was noted not only for its innovation, but also for the austerity of its Quaker founders. One worker from the mid-nineteenth century recalled 'primitive and paternalistic' conditions: 'The quiet of Union Street was even more marked between 9.00 to 9.20 when all employees attended a morning meeting. It was not uncommon to see passers stop to listen admiringly to the peaceful strains of a hymn sung by our girls and workmen as a prelude to the working day.'

As Richard and George Cadbury were struggling to establish their firm in Birmingham during the 1860s, according to *Fry's Works Magazine*, 'so great had become the expansion of our trade' that the factory was inadequate to deal with the 'orders pouring into the House from every quarter'. At a time when the Cadbury brothers were just beginning to recruit travellers to support Dixon Hadaway, Fry's travellers reached across England. George learned that a single Fry traveller with a flair for sales managed to secure ninety-five accounts in just four towns: Cheltenham, Stroud, Worcester and Gloucester. Gloucester alone bought £10,000 of Fry's goods. In the age of the steamship, the Frys also benefited from the Bristol docks that linked the company to Britain's burgeoning Empire and an ever-expanding horizon. To cap it all, they took advantage of Bristol being a leading naval base and won a contract to supply the British Navy – almost doubling their orders overnight. For the military, cocoa was valuable because it was easy to transport in tins, and was warm and filling for the troops.

From the Cadbury brothers' loss-making warehouse in Birmingham, the Frys must have appeared invincible. George knew he had a great deal to learn, and travelling with Francis James Fry gave him the chance to find out more about their latest pioneering inventions.

In 1847 the Fry brothers had introduced a novelty into the Victorian market. They had experimented with mixing their cocoa powder with its by-product, the excess cocoa fat. Whether by accident or design, they hit upon a way of blending the two ingredients with sugar to make a rich creamy paste. This concoction was then pressed into a mould and left to set. The result was the first solid chocolate bar in Britain. It was a breakthrough: a way of mass-producing a chocolate product that could be eaten instead of being consumed as a drink. This made chocolate portable, and turned it into a totally new kind of snack that could be carried on a railway journey or taken to work. They called it Chocolat Délicieux à Manger.

Fry's new product held no excitement for those with a really sweet tooth. It was bitter, coarse and heavy, and probably only of interest to the dedicated few who also possessed a strong jaw. Initially sales were slow. Undeterred, the Fry brothers had glimpsed a sweeter, more solid future. They set to work on more recipes for chocolate confectionery that could be produced in bulk. Secretly they experimented with a new kind of white minty cream. This was made by boiling sugar in an open pan, whipping it to an opaque creamy consistency and adding mint flavourings to give it a fresh taste. After the minty cream had cooled and been cut into sticks, these were dipped in luxurious dark chocolate. By 1853, Fry's frock-coated travellers were opening their sample cases to reveal a brand new product: Fry's delectable chocolate-coated Cream Sticks. Shopkeepers were amazed when they tasted the first chocolate confectionery produced on a factory scale; it was rich and Christmassy, a real treat. Better still, mass production meant that the price was significantly lower than that of hand-made confections.

The recipe proved to be a success, and within a few years the Cream Sticks were reformulated as a new type of chocolate bar. The chocolate for these 'morsels of delight', as they were called in Fry's literature, was formed into a thin, light paste. The mint cream was set in hundreds of tiny moulds and taken to covering rooms, where 'scores of young damsels' with chocolate trays coated the batches. In 1866 the first wagonloads of Fry's Chocolate Cream found their way to the grocers and sweet shops of Victorian Britain. Preliminary sales of Fry's minty chocolate sensation may have been modest, but there was growing interest – and not just from British customers.

French chocolatiers, who had long had a reputation for exquisite hand-made confections, were also exploring ways to produce them in bulk. Just outside Paris at his chocolate works on the River Marne in Noisiel, Émile Menier hit upon a process not dissimilar to Fry's. He had inherited his business from his father, a chemist, who had originally used cocoa sweetened with sugar as a coating for his pills. Émile Menier developed the cocoa side of his father's business, and by the mid-nineteenth century he had created a method for pressing dark chocolate into a mould. Eye-catchingly wrapped in chrome-yellow paper, it was the first solid chocolate bar manufactured in France, and it proved so successful that Menier's output quadrupled in ten years, reaching almost 2,500 tons in the mid-1860s, a quarter of the country's total output. Émile was able to invest more funds in his factory. Originally powered by a humble watermill, it was now equipped with shining new steam turbines, creating such a splendid spectacle that the locals called it 'the cathedral'. Much of Menier's chocolate was exported, and like many European manufacturers, he had his eye on the large populations of Britain's industrial towns. Soon he was in a position to open a factory in Southwark Street in London.

To improve the texture of his chocolate and increase his production, Menier needed extra cocoa butter, the fatty part of the bean. He found a supplier in Weesp, near Amsterdam, where a cocoa-making family firm was run by Coenraad van Houten. Van Houten had managed to solve a problem that had eluded everyone else: how to mechanise the separation of the fat content from the rest of the cocoa bean, which produced cocoa butter as a by-product. As a result his cocoa was purer and more refined than anything else on the market. Exactly how he achieved this was a trade secret, but there was no secret about his sales: his agents were acquiring customers in London, Edinburgh and Dublin.

As a regular visitor to London, George Cadbury could not fail to notice the new products that were becoming available: a purer form of cocoa made by the Dutch, and eating chocolate manufactured as solid bars in bulk. In the 1860s sales of eating-chocolate in England were modest – nothing compared to the established drinking-cocoa brands. Even so, like a flag planted on new territory beating against the wind, they pointed the way to unlock the potential inside the little cocoa bean.

George recognised that the Fry family was better placed than anyone else in Britain to take on the foreign competition. Although none of their cocoas matched the quality of van Houten's pure Dutch cocoa, Fry of Bristol was the cocoa metropolis of the world. Its four factories on Union Street, towering eight storeys high, seemed as secure as their granite and concrete exteriors. Just as this towering citadel dominated the town, so the bounty within dominated the market. The variety and sheer abundance of Fry's chocolate temptations put them in a class of their own. The Frys were indeed a beacon, a light to follow.

The Cadbury brothers' need to produce a popular product was becoming critical. Lacking the money to invest in the moulding machinery that was necessary to mass-produce such luxurious temptations as a chocolate bar, Richard and George struggled on producing cocoa as a drink mixed with questionable starches to absorb the fat. Their new products, Iceland Moss, Pearl Cocoa, Breakfast Cocoa and others, had failed to make an impact, and their losses continued to mount.

In response to yet another grim stocktaking, it fell to Richard to tackle the overdue accounts. 'We made the lowest class of goods,'

The public showed little appetite for cocoa mixed with lichen.

George wrote later, and consequently they had some of the 'least desirable custom', who were not always ready or willing to settle their debts: 'The small shopkeepers were constantly failing.' Some went under without paying, putting Cadbury at risk of going under as well.

The brothers had resolved, whatever happened, not to take on any liabilities that they could not meet, or to turn to their father John for additional funds. Not unusually for the time, their only sister, Maria, now in her late twenties, had postponed any thoughts of marriage to devote herself to caring for their father. Inevitably she provided a focal point for family news. Their oldest brother John, after a brief attempt at farming in the West Country, had made the bold decision to emigrate to Australia, and had sailed from the East India Docks in London on 17 December 1863 on the ninety-day journey for Brisbane. Their younger brother Edward was embarking on a home-decorating business, and the youngest, Henry, was still at school. Maria, gentle, patient and unassuming, provided encouragement with a maternal eye.

Richard and George continued to work relentlessly, spending long days on the road selling their cocoas to reluctant grocers and returning to the warehouse to pack the orders themselves if hands were lacking. The shortage of money was proving a strain at home. Richard's oldest son, Barrow, later recalled a family outing to Pebble Mill, 'where there was a pretty stream'. His mother, Elizabeth, suddenly felt unwell, but his parents elected 'to tramp all the way back again when his father would have given so much, had he been able, to take her home in a cab'.

The struggle was also taking its toll on the brothers' relationship. Richard and George often took very different views about how to proceed. With their offices next door to each other, they were careful to keep any disagreements private, debating their options until one had convinced the other of the best way forward. To their staff they appeared united – according to Elizabeth they won the nickname 'the Cheeryble Brothers', after the philanthropic twins in Dickens's *Nicholas Nickleby*.

All the brothers' industry and virtue made no difference. At the end of four years they were facing disaster. 'All my brother's money had disappeared,' George admitted. 'I had but 1,500 left – not having married.' There were insufficient funds left of their inheritance to develop a business that was in desperate need of capital for mechanisation. 'I was preparing to go out to the Himalayas as a tea planter,' said George. 'Richard was intending to be a surveyor.'

The Cadbury business was all but dead.

5

Absolutely Pure and Therefore Best

George Cadbury was considering one final reckless gamble. It would consume every last penny of his inheritance, but as long as he did not fall into debt or the great disgrace of bankruptcy, he felt it was a risk that had to be taken.

The more he learned about the Dutch manufacturer Coenraad van Houten, the more intrigued George became. Dixon Hadaway told him that van Houten's refined, de-fatted cocoa was now so popular it was on sale in regional centres like Leeds and Liverpool, as well as London. Gradually George began to realise that this was the model he should follow. Refined cocoa surely held the key to the future.

George discovered that the secret of van Houten's success lay with an invention he had developed with his father, Casparus, more than thirty years earlier. The van Houtens knew that established methods of boiling and skimming the bean resulted in an indigestible cocoa consisting of over 50 per cent cocoa butter. After experimenting with various designs of mechanical grinders and presses, they eventually perfected a hydraulic press that reduced the cocoa fat to less than 30 per cent. This meant that the often unappealing extras that had formerly been used to mop up the fat were no longer required. The result was a purer, smoother drink that tasted more like chocolate and less like potato flour.

The Dutch process remained secret, and no one in England, not even Fry, had discovered a way to manufacture a purer cocoa. George pondered this possible opportunity. Could this be the way to get ahead of their English rivals? Would the Dutch be prepared to sell the brothers their machine? If they had it, could they turn the factory around by using the cocoa butter produced by the refining process to create fancy chocolates like the French? Suddenly George could see a business future that made sense. Instead of using the bean to create one type of product, a fatty and adulterated cocoa, he could create two different products – pure cocoa and eating chocolate – using the most appropriate bit of the bean for each. A whole new set of possibilities opened up – if he could get only hold of van Houten's machine.

'I went off to Holland without knowing a word of Dutch,' said George, and 'saw the manufacturer with whom I had to talk entirely by signs and the dictionary'. George, plainly dressed, earnest, frustrated by the language barrier and absolutely sure that the odd-looking machine would save the Cadbury factory, was desperate to charm and persuade van Houten, and take home the prize.

Despite the difficulties in communication, van Houten succumbed, and a Dutch de-fatting machine was sold to the English Quaker gentleman. The records do not reveal the price that was agreed, but

Van Houten's cocoa press.

it is likely that it accounted for much of George's remaining funds of around £1,000. He made arrangements to ship the monstrous machine back to Bridge Street. It arrived by canal, and the sturdy cast-iron apparatus, a full ten feet high, was then laboriously transported to the chocolate works. Worse, it was very greedy: its giant hopper had to be fed with a large amount of beans. The brothers had to find a way to shift their new drinking cocoa in volume, and fast.

Their preparations were hit by a double tragedy in the family. In January 1866 their younger brother, twenty-two-year-old Edward, died unexpectedly after a short illness. When their older brother, thirty-two-year-old John, wrote home in May from Brisbane to express his grief at the loss, Maria was alarmed to see that his writing appeared unsteady, and the letter was unsigned. Doctors in Australia diagnosed him as suffering from 'colonial fever', a form of typhus. News of his death on 28 May followed almost immediately.

George Cadbury in 1866, when the business was close to failure.

Richard Cadbury with his son Barrow in 1866.

The unexpected loss of two brothers in such quick succession made Richard and George feel their responsibilities even more keenly than before. The survival of the family business, and any hopes of future prosperity, depended on this last throw of the dice.

In the coming months, they streamlined production of the new drink. By the autumn, Richard was ready to start designing the artwork for the packaging. At last, in the weeks before Christmas 1866, Cocoa Essence was launched.

It soon became apparent that there was a problem. Unlike the cocoa of their competitors, which went further because of the addition of cheap ingredients such as starch and flour, the pure Cocoa Essence was by far the most expensive cocoa drink on the market. The launch faltered. Customers were scarce. The strain on the brothers was beginning to exact a toll. 'It was an extremely hard struggle,' George admitted. 'We had ourselves to induce shopkeepers to stock our cocoa and induce the public to ask for it.' It looked as though the gamble had failed.

To the Frys, watching from Bristol, the Cadbury brothers' move hinted at desperation. Under the management of Francis Fry, the company's sales reached a staggering £102,747 in 1867. He continued to invest, expanding the premises in Union Street, and following the contract with the navy, Fry's workforce rose to two hundred.

In 1867, George and Richard made one last effort, exploiting something that other Quaker rivals spurned on principle: advertising. Plain Quakers like the Rowntrees in York believed that a business should be built on the quality and value of its goods. Nothing else should be needed if the product itself was honest. Advertising one's goods was like advertising oneself: abhorrent to a man of God. To Joseph Rowntree, proudly established as 'Master Grocer' in his shop in York, advertising seemed slightly shabby and unworthy, elevating promotion above the quality of the product. Even though he could see that his younger brother Henry's cocoa works at Tanners Moat was not taking off as hoped, he did not consider advertising to be the answer. He dismissed it as mere 'puffery'; he even objected to fancy packaging, and was content to alert his customers to any new product with a restrained and dignified letter. His deeply religious sensibility was offended by the idea of extravagant claims or exaggeration of any kind.

The Frys too had Quaker sensibilities when it came to excessive promotion. With the confidence that comes with over 150 years as a successful family business, Francis Fry saw little need for change. 'Our early advertisements had a certain coy primness about them,' conceded Fry's management in the company's 1928 *Bicentennial Issue*. The 'venerable announcements' of their original drink in the eighteenth century, Churchman's Chocolate, consisted of long-winded essays trying to explain why this product was unique and how to obtain it – by Penny Post or from 'the hands of errand boys'. This had progressed by the early nineteenth century to little homilies that advised the public how to prepare the drink and why it was good for them. But the language remained old-fashioned, describing the firm as an 'apothecary'. 'We were full of innocent pride in that period,' wrote the management. Certainly these notices had nothing in them that would stop readers in their tracks. No gorgeous girl of forthright demeanour with glossy lips and an unmistakeable message in her eye as she drank her cocoa: just a message, hardly readable in small typeface, telling of Churchman's Chocolate.

To the Cadbury brothers it seemed that advertising could do more. Another company, Pears, was taking advertising to new levels at this time. In 1862 Thomas Barratt, often described as the father of modern advertising, had married into the Pears family, and saw a way of turning a little-known quality product, Pears soap, into a household name by replacing the traditional understatement with a simple, attention-grabbing message. He began by enlisting the help of eminent medical men such as Sir Erasmus Wilson, President of the Royal College of Surgeons, and members of the Pharmaceutical Society of Great Britain. At a time when some soap products actually contained harmful ingredients such as arsenic, the medical men were happy to endorse Pears because it was 'without any of the objectionable qualities of the old soaps'. Barratt created posters and packaging adorned with eye-catching images of healthy children and beautiful women, with the brand name featuring boldly. For the consumer the message was immediate and simple: use this soap and you will be beautiful.

To the Frys, the modern style of poster, with its pithy message, was not unlike 'a sudden assault' on the eyes. 'You get the great news *at once*,' declared Fry company literature. 'You feel that something has struck you and you have, of course, *been* struck, not by somebody's fist or stick, but by an *idea*.' For many Quaker firms this 'assault' on the unsuspecting customer raised ethical concerns, but desperation drove the Cadbury brothers to a different view.

George and Richard knew they had to change the public perception of their pure new drink. Shrugging off their Quaker scruples, they took a gamble and committed to a further investment. Like the Pears team, they asked their salesmen to visit doctors in London with samples of their new product. To the delight of the brothers, this won the support of the obliging medical press. 'Cocoa treated thus will, we expect, prove to be one of the most nutritious, digestible and restorative of drinks,' enthused the *British Medical Journal*. Noting the brothers' claim that their product was three times the strength of ordinary cocoas and free from 'excess fatty matter', the *Lancet* concurred: 'Essence of Cocoa is just what it is declared to be by Messrs Cadbury brothers.'

The Cadburys' timing was excellent, because during the 1860s the purity of manufactured foods was a growing concern for the public. There was very little regulation of the food market, and even staples like bread could be contaminated. The public had first been warned in 1820, when the chemist Frederick Accum published *A Treatise on Adulterations of Food and Culinary Poisons*, which argued that processed food could be dangerous. By the 1850s, Dr Arthur Hassall had written a series of reports in the *Lancet* exposing some of the commonly used additives in cocoa production: brick dust, red lead and iron compounds to add colour; animals fats or starches such as corn, tapioca or potato flour to add bulk. By 1860, in response to public pressure, the government introduced the first regulations to prevent the adulteration of food.

Yet still the practice continued. In one government investigation more than half the cocoa samples tested were found to be contaminated with red ochre from brick dust. Consumer guides appeared, telling customers how to test their cocoa and warning that a slimy texture and a cheesy or rancid taste indicated the presence of animal fat. If the cocoa thickened in hot water or milk this was evidence that starches had been added, something you could confirm if your comfort drink turned blue in the presence of iodine. Most worrying of all was the continued use of contaminants, including poisons such as red lead, which were injurious to human health but which enhanced the product's colour or texture. It was small wonder then that the *Grocer* hurried to follow the lead of the medical press and sang the praises of the Cadbury brothers' pure new product: 'There will be thousands of shop keepers who will be glad of an opportunity to retail cocoa guaranteed to contain nothing but the natural constituents of the bean.'

With this support, in 1867 the Cadburys planned the largest advertising campaign they had ever undertaken. There was no longer a question mark over advertising. They would use it with confidence, and really make the Cadbury name stand out. For the first time, they were effectively rebranding the whole image of cocoa. Their Cocoa Essence was honest, and they intended to shout it from the rooftops. Richard Cadbury came up with a slogan that capitalised on the strengths of their new product: 'Absolutely Pure, Therefore Best'. They took out full-page advertisements in newspapers and put posters in shop fronts and even on London omnibuses. Soon the Cadbury name, synonymous with the purity of the company's product, was everywhere. It was unavoidable, rippling around the capital like a refrain from a

CADBURY'S COCOA
ABSOLUTELY PURE, THEREFORE BEST.
NO CHEMICALS USED.

song. Given half a chance, the Cadbury brothers would have covered the dome of St Paul's, protested one writer. But at the chocolate works, everyone caught the mood of excitement.

By the autumn of 1868, with the campaign gaining momentum, the staff at Bridge Street grew to almost fifty. David Jones, a former railway goods porter who had longed to be a traveller, vividly recalled his first day: 'George put a sample in my hand and told me to go wherever I wanted for a week, the only stipulation being that I should

Horse-drawn omnibuses carried the Cadbury brothers' first poster campaign.

not trespass on the grounds of another traveller.' He chose north Wales, but soon had reason to regret his decision. No one had tasted anything like Cocoa Essence before. 'I gave hundreds of shop keepers a taste,' he remembered, 'only to watch their faces lose their customary shape as though they had taken vinegar or wood worm.' But Jones would not give up. He managed to secure thirty-five orders, and was gratified when the Cadbury brothers were 'highly pleased'. Many years later another traveller, John Penberthy, would also remember the thrill of winning orders: 'The delight of travelling in those ancient days, working towns not previously visited by a Cadbury traveller, surpassed in my opinion . . . the discoveries of Shackleton, Peary or Dr Cook!'

While pressing on with the launch of Cocoa Essence, the Cadbury brothers also followed Fry's lead with experimental types of eating chocolate. Their father, John Cadbury, had tested out a French eating chocolate before, but now that they had a large volume of creamy cocoa butter as a by-product of their pure cocoa drink they could dramatically increase the manufacture of eating chocolate. Rather than mimic Fry's rough chocolate bar, Richard and George wanted something altogether more luxurious. They found that when the excess cocoa butter was mixed with sugar and then cocoa liquor was folded back into the mix, it produced a superior dark chocolate bar. Then they went one step further. They wanted to launch a new concept that they hoped would bring the exotic products of the French chocolatier to the popular market. Richard called it the Fancy Box.

Had the Cadbury brothers not been in charge of a chocolate factory that was still faltering slightly, the lavish contents of the Fancy Box would undoubtedly have violated their principles. It represented the most un-Quakerly immoderation and extravagance. Generations of Quakers before them had maintained a beady-eyed vigilance in the pursuit of 'truth and plainness'. The senses on no account were to be indulged; the path to God demanded a numbing restraint and self-denial. But Richard and George, the apparently devout Quakers, had come up with the ultimate in wanton and idle pleasure. For each Fancy Box was a sensual delight.

The lid opened to release the richest of scents, the chocolate fumes inviting the recipient with overwhelming urgency to trifle among the luxurious contents as a whiff of almond marzipan, a hint of orange, rich chocolate truffle, strawberries from a June garden encrusted with thick chocolate beguiled the very air, all begging to be crushed between tongue and palate. Each one had a French-sounding name, adding yet more forbidden naughtiness: Chocolat du Mexique, Chocolat des Délices aux Fruits, and more.

It is ironic that George and Richard dreamed up these chocolate indulgences at a point when their own lives had become most Spartan. 'At that time I was spending about 25 pounds a year for travelling, clothes, charities and everything else,' George wrote. 'My brother had married, and at the end of five years he only had 150 pounds. If I had married, there would have been no Bournville today, it was just the money I saved by living so sparely that carried us over the crisis.' It is arguable that the brothers' unremitting self-denial fuelled their appreciation of sensual extravagance.

In the pursuit of plainness, Quakers spurned most artistic endeavour as a worldly distraction that could divert them from the inner calm that led to God. As a result George and Richard's father never allowed a piano in the house, and had given up learning his treasured flute. Any form of aesthetic enjoyment, such as theatre or reading novels, was discouraged; only texts of a suitably thoughtful tone such as Bunyan's *Pilgrim's Progress* or Foxe's *Book of Martyrs* were acceptable.

As for painting, this was considered a superfluous indulgence that could lead a Quaker astray with a false appreciation of something 'worthless and base'. But now Richard overturned these rigid rules. Revelling in exuberant splashes of colour, he began a series of paintings himself to be pasted on the covers of the Fancy Box.

Richard had travelled to Switzerland and made sketches of the Alpine scenery. Now these drawings, along with images of the seaside and even his own children, formed the basis of his designs, which were chosen to appeal to Victorian sentimentality. His own young daughter Jessie loved to pose with her favourite kitten. 'Among the pictorial novelties introduced to the trade this season, few if any excel the illustration on Messrs Cadbury's four ounce box of chocolate crèmes,' enthused the *Birmingham Gazette* on 8 January 1869. 'It is chaste yet simple, and consists of a blue eyed maiden some six summers old, neatly dressed in a muslin frock, trimmed with lace, nursing a cat.' To strike a real note of luxury, Richard decided that some of the Fancy Boxes should be covered in velvet, lined with silk, and include a mirror. In every way, Cadbury's chocolate was to stand for quality. The reviewer writing for the *Chemist and Druggist* magazine of 15 December 1870 was certainly won over. 'Divine,' he declared. 'The most exquisite chocolate ever to come under our notice.'

It was one thing to dream up recipes for the Fancy Box, but quite another to mass-produce them. 'When I think how we were cramped up in small rooms at Bridge Street,' recalled Bertha Fackrell of the Top Cream room, 'the wonder is to me now that we turned out the work as well as we did.' A lack of space was the least of their problems. 'Oh the job we had to cool the work!' Bertha continued. Although there were small cupboards with ventilators around the room, all too often when staff from the box room came to collect the crèmes and chocolate balls they were still too warm. 'I remember once we girls put our work on the window sill to cool when someone accidentally knocked the whole lot down into the yard below.'

Sales of the Fancy Boxes increased, and gradually more staff were hired. One new worker, the crème beater T.J. O'Brien, was amazed to find the owners grafting with the workers. 'During these trying times I never knew men to work harder than our masters who indeed were more like fathers to us,' he wrote. 'Sometimes they were working in the manufactory, then packing in the warehouse, then again all over the country getting orders.' O'Brien's work beating the crèmes was heavy, and 'often Mr George and Mr Richard would come and give me a help'.

But for all their hard work, reward was not to come easily to the Cadbury brothers. For Richard, busy pouring all his energy into the factory, the enjoyment of success, so longed for, so hard won, was wiped away. His adored wife Elizabeth died at Christmastime in 1868, ten days after giving birth to their fourth child. Suddenly his achievements seemed as nothing. The very centre of his family was gone.

At thirty-three, Richard was left with four very young children. Barrow was the oldest at six, followed by Jessie, who was three, one-year-old William and the new baby, named after Richard. 'He was everything to our baby lives,' said Jessie of her father. 'I can well remember riding on his shoulders and going to him with all our troubles.' However pressed Richard was at work she recalled, 'He was so much to us always.' The loss of his wife, in a Quaker household, required 'humble submission to God's will'. The children learned fortitude from their father. For Jessie, the certainty of her father's love made her feel 'it was worth braving anything'.

Perhaps because Richard grew much closer to his children after their mother's death, during the spring of 1869 he found the time to set up a crèche for poor or abandoned children and infants in the neighbourhood, renting a house for the purpose and enlisting the help of a friend, the motherly and highly competent Emma Wilson. Mrs Wilson had been widowed seven years earlier, and had managed to earn an income and raise seven children on her own. She became indispensable, not only in the nursery but also by helping out with Richard's children at their home in Wheeley's Road.

Sometimes Richard's children accompanied him to his office. Barrow had a vivid memory of going to Bridge Street with his father, and delighting in watching boxes being unloaded from the colonies. 'One day a large boa constrictor emerged and was chased by two men who held it down with sugar and cocoa bags,' he recalled. 'It was a revelation that the boa constrictor could bend its body with such force whatever the strain.' When the frightened boy fled to the Cocoa Essence sieving room, he was soon discovered, 'and given a lecture on the impropriety of being there'. Hygiene was all-important; no leniency was given even when hiding from a boa constrictor.

Although the Cadbury brothers' financial position had improved, they could not yet feel secure. None of their capital remained. Their livelihood and their future depended on the public buying their confections, charmed by an image of a blonde, blue-eyed girl holding a kitten and smiling sweetly from the lid of a chocolate box.

Richard's illustration of his daughter Jessie and her kitten.

In York, the Rowntrees, with a quality grocer's shop in the centre of town and a big chocolate factory by the river, appeared to be enjoying enviable success. But Joseph Rowntree, successful purveyor of superior foods, was worried. The problem lay with his brother Henry, and the chocolate factory at Tanners Moat, which looked more like a medieval castle with its forbidding high walls and blackened windows. It was becoming apparent that Henry had optimistically overreached himself by investing in the rambling complex on the River Ouse. By 1869, after a seven-year battle, his cocoa was still struggling to find a market, and the firm's future looked uncertain. The prospect of failure was very real.

The lesson of years of stern homilies about honouring debt, plain dealing and trustworthiness began to preoccupy Joseph. He knew perfectly well from the family's much-thumbed copy of the *Rules of Discipline* that good Quakers should keep 'a watchful eye over all their members and those heading for commercial trouble should be warned and if required, helped in their difficulties'. Joseph knew that Henry was in trouble, running his business in a most eccentric manner, with

equally eccentric accounting methods. Their father had died shortly
after setting Henry up in the cocoa business, and as his older brother,
Joseph felt keenly aware of his duty. Much though he delighted in
his role as 'Master Grocer', he could not allow himself the easy path
of putting self-interest first when his brother was in need of help.

At thirty-three, Joseph Rowntree had already won a reputation as
a man who took his Quaker responsibilities seriously. Having witnessed
the horror of the Irish Potato Famine, he had made the time while

Joseph Rowntree in 1862.

running his grocery shop to undertake an exhaustive study of poverty
in England. Adopting a true spirit of enquiry, he had attempted to
investigate not just the effects but also the causes of poverty.
Researching back to the time of the Black Death in the fourteenth
century, he had carefully gathered facts on pauperism, illiteracy, crime
and education, leading him to uncover a complex web of connections

that could trap a family in poverty. Concerned at the apparent indifference of the authorities, he wrote up his findings as a paper, *British Civilisation*, which he hoped to present at an Adult School Conference in Bristol in 1864. But even fellow Quakers, such as Francis Fry's nephew Joseph Storrs Fry II, who was running the conference, feared his essay might 'cause offence to weak brethren', and urged him to modify its strong language. The following year Joseph Rowntree published a more measured paper, *Pauperism in England and Wales*, a landmark study that set out the figures and questioned the role of Church and state in perpetuating social injustice.

Rowntree's studies had given him confidence in his ability to collect data and analyse problems – skills that he reasoned could help him stabilise his brother's business. In 1868 he took a bold step, withdrawing his inheritance from the security of the Pavement grocery shop to invest with Henry, hoping to bring order to the chaotic chocolate factory by the river, where Henry was put in charge of production. There was every reason to believe that, properly run, the factory could make headway. After all, other firms were turning a profit from cocoa.

Joseph, who had a keen eye for detail, found much to vex him as he embarked upon a painstaking examination of his brother's accounts. Henry liked informality, and a number of most irregular practices and unbusinesslike anomalies had blossomed undisturbed in his factory. It seemed that each room, each account book or order book, each pile of receipts, harboured potentially fatal flaws. Joseph was confronted by a parrot in the workroom, and an obstinate donkey with a predilection for steam baths. The parrot distracted the workers, and the donkey failed to meet Joseph's exacting requirements for the firm's transport, stubbornly refusing to budge from the warmth of the steam pipes that emerged everywhere from the walls of cottages and outhouses that had been converted into factory buildings. The donkey had to go – to be replaced by a much more versatile handcart.

As for the accounts, Joseph's detailed personal notebooks from that time are filled with long columns as he tried to get to the bottom of the debts to York Glass, York Gas, the saddler and the parcel delivery service. To resolve discrepancies in the accounts, he was obliged to resort to hearsay to work out the company's liabilities: 'Beaumont says he *thinks* Epps gave a 7% discount upon his lowest whole sale quotation.' Henry's staff, like Henry himself, it was clear, were a trifle hazy when it came to the details of the deals they had made. Perhaps that was not surprising, as the staff that had been pared down to the bone. Seven workers managed the key processes of grinding, roasting, rubbing and carrying sacks from the warehouse, each taking turns with heavy work. There were definitely no spare funds to squander on something that Joseph regarded as disreputable as advertising.

As a small antidote to the Mad Hatter logic of the castle, Joseph Rowntree did cast a discerning eye over the competition. Recognising Cadbury's potential breakthrough with Cocoa Essence, he began to make discreet enquiries as to where he could purchase machinery to make a purer form of cocoa

Relaxed in the knowledge that he was at the helm of the world's largest chocolate company, Francis Fry found he could delegate the many day-to-day problems involved in running such an establishment to others, and devote time to his numerous philanthropic and public causes. For years he had had a keen interest in the West Country railways, and during 1867 he led a campaign to unite the nineteen separate western lines. In addition, it was his great ambition to help create a national Parcel Post which could carry parcels at uniform rates throughout the whole of Britain. On top of this, he took a particular interest in the Bristol Water Works, a scheme to replace the old city wells.

But Francis Fry had another little interest that consumed a great deal of his time while he assumed the factory was running along nicely: he wanted to create a definitive history of the Bible, nothing less than 'a systematic and historical account of the various editions of the different translators'. His profound conviction that the 'Sacred Scriptures were of Divine origin and he was unravelling the wishes of the Lord' necessitated extensive trips abroad retracing the footsteps of great Protestant scholars such as William Tyndale. In the early sixteenth century Tyndale was the first to translate much of the Bible into English directly from original Hebrew and Greek texts. He encountered so much opposition that he was forced to flee to Europe, where he was betrayed in 1535 and burned at the stake the following year.

Fry's passion for William Tyndale 'amounted almost to veneration', wrote his son and biographer Theodore. He viewed him as one of the 'greatest men England ever produced', and was determined to make an exact copy of the Tyndale testament of 1525–26, the first English Bible. In 1867, the year Cocoa Essence was advertised on London omnibuses and the Cadbury name was emblazoned all over London, Francis Fry was absorbed in publishing a treatise on the Tyndale testament. Not content with this, he also tried to track down the original Bible translations of Thomas Cranmer, the sixteenth-century Archbishop of Canterbury. Three years after the martyrdom of Tyndale, Henry VIII had authorised an English translation to be distributed to every church in the land, so that anyone could hear the word of God.

Little by little, Francis Fry's real world was reduced to a shadowy decade in the sixteenth century, entwined in the internal problems of the Tudor court. His factory, the great chocolate citadel in Bristol, was a faded dream compared to his glorious quest to find the chalice, the authentic word of God. While he was so intensely preoccupied, it was hard for him to pay more than fleeting attention to the modernising force being pioneered by Cadbury.

At Bridge Street, Richard and George Cadbury were beginning to find their wilderness years were behind them. The first signs, reported Thomas Little in the packing room, came from a traveller in the Black Country: 'The weight of the goods had broken the springs of his van, and he had had to run it into a customer's cart house for repair and ride home on a horse.' This was one of many clues that the company was turning the corner. Skilful use of technology and advertising were winning customers. And once the Cadbury brothers knew what they were doing, the ideas kept coming. Records for the Birmingham patent office show that on 3 November 1869 they sealed a patent for one of the first kinds of chocolate biscuit – 'a new improved description of biscuit manufactured from the cocoa bean' – and the search was on for other new forms of luxury foods.

All this was not enough for George. Seizing the initiative from his rivals, he went on the offensive to promote Cadbury's pure new cocoa. Aware of the public's growing sensitivity to food adulteration, he lobbied the government to take action. People should know what they were buying, he argued. Eventually he was summoned to a government committee to give evidence. In a troubling move for his competitors, he insisted that only an absolutely pure product such as Cadbury's Cocoa Essence should be called 'cocoa'. All preparations mixed with additional ingredients should be sold under a different name.

The Frys and other cocoa manufacturers finally woke up to the threat. They too began to promote their products, claiming that they only used nutritious additives. It was a bitter battle in which Cadbury benefited greatly from the free publicity, much of it at the expense of its rivals who were protesting against the proposed new regulations.

The victory finally went to George Cadbury when the government introduced the Adulteration of Food Acts in 1872 and 1875. Under the new legislation, *all* ingredients in cocoa had to be listed. This meant that the public could now see for themselves that Cadbury's Cocoa Essence was the purist form of cocoa. Grocers who stocked adulterated cocoas without proper labelling could be prosecuted. Records show that the Marylebone police prosecuted a Master Grocer by the name of Mr Kirby of 212 High Street in Camden after he sold an inspector two samples of cocoa that were adulterated: one 'was manufactured by Messrs Taylor Bros . . . and the other by Messrs Dunn and Hewett and Co.'.

As sales of Cocoa Essence rocketed in the early 1870s, the number of employees at the Bridge Street works grew rapidly. For Richard and George, the view from their office windows formed a stark contrast to the outlook of ten years ago. Then, an unmistakeable air of neglect had surrounded the apparently dying firm. Now, all was bustle and busyness, and the applause of horses' hooves from the crowded courtyard where carts and carriages waited to transport stock.

The early 1870s were markedly different for the upright Joseph Rowntree and his brother Henry: their company was still struggling to survive. Joseph meticulously pasted into his notebook some Cadbury flyers that were full of exasperating claims: Cadbury's cocoa went 'three times as far' as the best of the adulterated cocoas, and just one halfpenny would 'secure a delicious cup of breakfast cocoa'. But Joseph was not in a position to take them on. Quite the reverse. He was having such a struggle that even the passionate and indignant author of *Pauperism in England and Wales* found himself able to relax his Quakerly ideals just a little. Rowntree needed a shortcut, a little knowhow, some real expertise in turning the unlovely-looking cocoa bean into something deliciously edible in an enticing box.

Joseph's private notebook reveals that in March 1872 he went to London to recruit new employees, and did not see anything wrong in conducting a little industrial espionage at the same time. He rented premises at 314 Camden Road in central London and placed advertisements in the capital's papers, some in the vicinity of Taylor's Spitalfields works:

To
COCOA AND CHOCOLATE MAKERS
WANTED IMMEDIATELY
A FOREMAN *who thoroughly*
understands the manufacture
of Rock & other Cocoas, Con-
fection and other Chocolate
*

Also several WORKMEN *used*
to the trade
good hands will be liberally dealt
with

Aware of the need for discretion, the Rowntree name did not appear on the advertisements. Applicants were advised to 'Apply by letter only to: "G.F., 12 Bishopsgate St EC".'

Recording every detail in his elegant longhand, Joseph Rowntree found that the Taylor brothers' employees were a willing mine of information. He soon received eager replies from a number of their workers, such as mixer and foreman James French:

Gentlemen
In answer to an Advertisement in the Clerkenwell News, I beg leave
most respectfully to offer my services as a Mixer, having been in
Messrs Taylor Bros factory for Two years, and understand making

Rock Cocoa and others, but not Confection, wishing to better myself.
The favour of a reply will be immediately attended to by Gentlemen
Your obedient servant
James French

Joseph conducted interviews between Thursday, 7 March and
Tuesday, 12 March. His records reveal that the interviews went far
beyond discussions of the applicants' experience. Workers were willing
to discuss every aspect of the Taylors' business, including details of
manufacturing and even precious recipes – for a sum. One of the first
men Joseph hired was James French, who had his fare to York paid,
was offered twenty shillings a week and, most important, given a
welcoming gift of £5 for his Taylor recipes.

French introduced Joseph to other workers, such as Robert Pearce
of Whitechapel who claimed to manufacture 'all of Taylor's Chocolate,
Chocolate Sticks, and Confection Chocolate'. Pearce was duly invited
to come to the York factory during his three-week holiday, when he
would receive £2 a week 'for imparting all his knowledge'. Others
followed. William Garrett, who had been at Taylor's for twelve years,
received two shillings and sixpence for, among other things, a recipe
for 'Unsworth's Cream Cocoa'. Joseph learned about the technology
used by the Taylor brothers from Henry Watkins, of Waterloo Town
East, a man described as 'the cleverest man on Taylor's Flake Floor',
who could 'take a mill to pieces and put it together again'. And after
meeting with James Mead, also of Taylor's Flake Floor, who received
ten shillings, Joseph was able to set out pages of detail covering the
company's manufacturing processes. He now knew the exact order
in which the Taylors added the ingredients for Rock Cocoa, the ratios
of different types of bean and even the temperature of the mixer: 'as
hot as the hand can comfortably bear and sometimes [by accident]
gets hotter'.

Quietly, almost without their noticing, Taylor's men were relieved
of valuable expertise accumulated over years of exacting work in the
many different processes involved in the making of chocolate. All
their labour and expertise had been fed to, gobbled up and digested
by the dark, religious man with the worried look. For a modest outlay,
Joseph Rowntree was now possessed of enough information to dupli-
cate key processes in Taylor's factory.

It seems that Joseph was not overly troubled by his conscience, for in April he was back in London to obtain more information, before a brief trip to Germany where he gathered price lists and technical information about companies such as the Stollwerck brothers of Cologne. June brought him back to London. In the interim more advertisements had appeared in the capital's local papers, such as the following in Stoke Newington. Having gleaned all he could about Taylor's Rock Cocoa on his first visit, this time he wanted to learn about Soluble Cocoa.

WANTED

MEN *Who thoroughly under-*
stand the Manufacture of
SOLUBLE COCOAS

Apply by letter, to HH., 19 Lordship Road, Stoke Newington, N

Henry Richard Thompson of Islington, who had been at Dunn and Hewett in Pentonville for thirty years, was offered an opportunity 'to come down for at least 4 weeks to teach all he knows, wages 2 pound per week. One pound to be allowed for each railway journey to York and a lump sum of 10 pounds to be given for the receipts and the knowledge'.

Joseph also decided to take a closer look at some of his Quaker rivals. He caught the train to Bristol and met some of Fry's workers at an address in Elton Terrace in Bedminster, including J. Charles Hanks, who claimed to have all Fry's 'fancy goods' recipes. In Birmingham, Joseph was particularly keen to make contact with French workers involved in the manufacture of Cadbury's Fancy Box. Soon after, he went to Paris. The object of this journey, he wrote, was to make enquiries of certain French chocolatiers, including Émile Menier at Noisiel sur Marne.

It speaks volumes about the struggle to survive the intense competition that Joseph Rowntree, a man who in public epitomised Quaker virtue, and whose principles led him to spurn advertising, should engage in such subterfuge. Bribing workers to disclose their employers' secrets fell far short of the ideals of honesty and plain dealing required of Quakers, but if Joseph wrestled with his conscience, there is no record of it in his notes. Remarkably, even as he was engaged in this discreet espionage, he also wrote formally to both Fry and Cadbury to suggest that they unite on price and discounts to help them deal with the foreign competition. Such unified action, he reasoned, would enable Quaker values to survive.

While Joseph Rowntree was obliged to compromise his ideals in his attempts to stay in the game, there was a far worse catastrophe in store for another chocolate manufacturer in the early 1870s. The Taylor brothers of Spitalfields, second only to Fry among the English manufacturers, saw their fortunes change dramatically in just a few hours.

Walking through Spitalfields in the small hours, some young men saw smoke billowing from Taylor's cocoa, chicory and mustard factory. One rushed to inform the police, but just 'minutes afterwards, flames were seen to burst from the windows of the basement floor', according to the local press. Almost immediately the conflagration erupted with a roar and the whole building was outlined in fire, glowing like a Roman candle as sparks shot high into the night sky, feeding on the factory's incipient warmth and fatty, inflammable contents.

Such was the scale of the emergency that engines from Bishopsgate, Watling Street, Farringdon Street and other fire stations were soon on the spot. 'No time was lost in commencing to play upon the buildings,' but despite the efforts of the firemen, 'floor after floor became ignited in rapid succession'. The Taylor brothers arrived to find an inferno. More than five separate buildings, one eight floors high, had become 'one mass of fire': 'ultimately the roofs gave way and fell with a loud crash; the flames at this time being so brilliant as to illuminate the metropolis for miles around'.

By daylight, Messrs H. and J. Taylor found their premises completely gutted, only two floors of one of the warehouses remaining intact. One of the largest firms in the kingdom had been reduced to a pile of ash. Fortunately, according to the press, it was understood that the loss would be covered by the company's insurance.

After the introduction of the new legislation, orders began pouring in for Cadbury's Cocoa Essence. The factory, for so long a millstone that had imprisoned Richard and George in a rigid regime of work, was in profit, and its success began to impact on their private lives. Richard was ready for some warmth and charm in his life. He met and fell in love with Emma, the daughter of Mrs Wilson, who ran the crèche, and bought a home that backed onto the canal. 'Every time I come into the house I think of you,' he wrote to Emma. 'It seems like one real step to having you here, to have a home for you . . . I have given you all my heart, and I have not much else to give you, but all that I have seems to belong to you quite as much as to me.' They married in July 1871.

Even George, whose horizons had been so narrowed by work, fell under the spell of their modest success. He met twenty-two-year-old Mary Tyler through a cousin, and a relationship soon developed. Ten years of ruthless self-denial and austerity made it hard for him to express his emotions. 'A Spartan severity,' writes his biographer Alfred Gardiner, 'was the key note and the senses were kept in rigorous and watchful restraint.' The reserve of his letter of proposal contrasts with that of his brother: 'I feel that thou dost love the Saviour,' he wrote to his prospective fiancée, 'and that if we were united together it would be in Him, and that thus united we should calmly, peacefully & joyously pass through life's journey.'

Mary was so confused by George's formality and unromantic approach that she consulted her mother, who was equally baffled. 'We are quite at a loss how to counsel thee . . . It certainly struck us the letter was written without ardour, and in a business like manner,' Mrs Tyler replied to her daughter, 'without even saying that he felt a strong preference for thee.'

George may not have realised just how much his asceticism and restraint, which had served him so well in business, threatened this delicate opportunity. His prospective relationship hung in the balance. Mr and Mrs Tyler sensed that their daughter was unsure. 'Are we not right in judging that thy feelings on the subject are a little doubtful and mingled?' enquired her mother. She went on, 'if looking to the future thou feels pretty sure thou couldst not really enjoy his companionship in the very nearest of relations, why the best way is to send him a positive refusal'. And, she added, to do so 'as soon as thou well canst'.

Mary could not bring herself to this point, and later that summer her parents arranged to meet George for a short break in Southend. Away from the confines of the factory, George's feelings finally became evident. Abandoning his ingrained sense of discipline and restraint, when he arrived in the town and found that Mary was not at the arranged meeting place, to the evident of relief of her parents 'he set off to run like a boy, running all the way to our lodgings' to find her. Over the following few days the typically reserved and saintly George found a way to express his love. Friends noted Mary's 'sweet look of quiet joy', and found it 'delightful to see his love for Mary and hers for him'.

The couple married in April 1872. An extensive honeymoon in Switzerland, France and Rome was planned. George abandoned the shackles of business for the first time in more than a decade for this tour of Europe with his young bride. As he boarded the train for France, the world of the factory faded: but not much.

PART TWO

6

Chocolate that Melts in the Mouth

Unknown to the Cadbury brothers, two Swiss entrepreneurs were secretly working on a breakthrough that would transform the destiny of the 'food of the gods'. In doing so, they had the potential to destroy the English chocolate manufacturers.

The legend began in a small way in the 1850s, when a young entrepreneur, Daniel Peter, completed his apprenticeship with a candle-maker in Alsace and moved to the picturesque town of Vevey, nestled in the Swiss Alps. But his plan to set up shop as a candle-maker with his brother Julian was overtaken by events. The discovery of a method of distilling kerosene from oil was swiftly followed by the development of the kerosene lamp, the clean-burning light of which made the flickering tallow candles and whale-oil lamps of the past obsolete. The future looked lighter and brighter, but not for a candle-maker.

Daniel Peter continued to run the Frères Peter candle company with his brother, but he also found time to pursue his keen interest in the manufacture of food. With breakthroughs in food processing, tinned meat, soup and bottled fruit were just some of the novelties now being dished up by man and machine. The centuries-old stale ship's biscuit was being replaced by miraculous new temptations; even condiments were getting a modern makeover. The days when convenience food tended to be soup or hot eels bought from a street vendor were over. 'Without a doubt, industrial products intended as food offer manufacturers the best prospects for success,' observed Peter. 'These processed foods are consumed every day and unlike other products, there is a constant demand for them that is not subject to the whims of fashion.'

Peter's opportunity to move into the food business proved to be right on his doorstep. In 1863 he married a local girl, Fanny Cailler, whose father's chocolate factory was the first in Switzerland to mechanise the process of grinding cocoa beans. For Daniel Peter, his father-in-law's chocolate business was an inspiration. The more he learned about cocoa, the more convinced he was that it had an exciting future. Cocoa, he predicted, would become a regular part of people's diet, like coffee. Full of optimism, he set off to Lyon to work in a chocolate factory and master the exotic craft of the French chocolatier.

In 1867, the year the Cadbury brothers' Cocoa Essence was taking off in Britain, thirty-one-year-old Peter returned to Vevey to start a chocolate business of his own. He and Fanny settled at 13, rue des Bosquets, and next door at number 12 he proudly attached the nameplate of his second company: Peter Cailler et Compagnie. In this picturesque setting, nestled in the foothills of snowcapped mountains, he hoped to create the perfect chocolate.

According to one family story, it took a crisis to lead Peter to the breakthrough that would transform chocolate across the world. On 30 September 1867 Fanny gave birth to a baby daughter, Rose Georgina. But there was a problem – Rose rejected her mother's breastmilk. With each hour slipping by and their baby unable to feed, the young parents became distraught. Peter appealed for help from a

Daniel Peter's chocolate works at Vevey, Switzerland, c.1867.

neighbour who lived a few doors away at 17, rue des Bosquets. The man he turned to was none other than the acclaimed German inventor Henri Nestlé.

Henri Nestlé looks out from his sepia nineteenth-century photograph, his dark, deep-set eyes emphasising his intense expression. His thinning hair is neatly swept back from a broad, well-shaped forehead; his beard, just a little unruly, provides the only hint of disorder. This imposing figure was known locally as a 'merchant', but he had real flair as a scientist and entrepreneur.

At the time Daniel Peter arrived on his doorstep looking for help,

Henri Nestlé, c.1875.

Nestlé was on the verge of a life-changing advance. He had just begun selling a special type of 'milk flour' for babies, using his own formula for powdered milk. Peter and his wife were desperate for baby Rose to try Nestlé's special formula.

Born in Frankfurt in 1814, Heinrich Nestle left Germany as a young man, and like Peter he chose to settle by the beautiful shores of Lake Geneva in Vevey. Changing his name to Henri Nestlé, he rapidly demonstrated his versatility as an entrepreneur, a maverick and a scientist. Quite apart from his trade as a druggist, selling medicines, seeds and mustard, his interest in oil lamps led to a flourishing small business manufacturing liquid gas. Like Peter, Nestlé was intrigued by developments in food manufacture, and by 1847 he had begun to research infant feeding. This was an era plagued by infant mortality: in Switzerland, one in five babies died before their first birthday. The challenge Henri Nestlé took up was to create a new type of food for babies whose mothers were unable to breastfeed.

Since milk turned rancid so quickly, the puzzle was how to keep it fresh. Using his own kitchen as a laboratory, Nestlé experimented with different methods of preserving whole milk. By 1866 he had a solution: he found a way to create a milk powder concentrated by an air pump at low temperature that he claimed would remain as 'fresh and wholesome' as if it was 'straight from the cow's udder'. To this he added a cereal, 'baked by a special process of my invention', to create a unique formula: *'farine lactée'*.

By chance, in September 1867 Henri Nestlé was approached by a local doctor friend who was treating a premature baby. The child was convulsive, and could not breastfeed or keep down any alternative; his mother was very ill as well. After fifteen days, the baby's survival hung in the balance. But to the family's delight, he was able to digest Nestlé's formula. News of this 'miracle' spread across town.

It was an anxious moment when Daniel and Fanny Peter tried to coax baby Rose to drink Henri Nestlé's formula. The child was fretful, hungry but unable to keep food down. After a few moments came the sounds the distraught parents had so longed to hear: normal suckling from a contented baby. Rose's health soon improved, and she began to put on weight.

Doctors tested the product on other infants, with the same result. Despite sceptics who insisted it was no more than a 'sack of flour', Nestlé was full of confidence in his new invention. 'My discovery has tremendous value,' he declared, 'for there is not another food comparable to my baby food.' In 1868, after successful launches in Vevey and Lausanne, and his home town of Frankfurt, demand continued to rise. Nestlé did not do things by halves. He employed a sales team in France, and ventured to England to open an office in London. 'Believe me it's no small matter to market an invention in four countries simultaneously,' he said. He installed a large new vacuum pump, and was able to manufacture more than a half ton of dried infant milk a day. Even with the outbreak of the Franco-Prussian War in July 1870, which made it harder to move goods around Europe, the company's growth seemed unstoppable.

Through his friendship with Henri Nestlé, Daniel Peter began to see an opportunity. It occurred to him that his chocolate products and Nestlé's technology for creating powdered milk might be combined to create a creamy *milk* chocolate drink. After all, people were already accustomed to mixing cocoa powder with milk to make a drink. Why not blend the two to create a ready-made milk chocolate powder? And if he succeeded, perhaps he could adapt his formula for pre-combining cocoa and milk into other convenient and luxurious chocolate treats.

From his small warehouse by the lakeside, Daniel Peter experimented with adding Nestlé's dried milk to cocoa and sugar. After his initial optimism, he hit problems: when cocoa was combined with the milk powder, the resulting drink was coarse and grainy; but when he processed the milk himself, the water in it did not blend well with the oil in the cocoa bean. In addition, the water reacted with the sugar to alter the texture. Somehow he had to find a way to dry the milk without spoiling it.

Peter became obsessed with the challenge. His plant was open around the clock: during the day he made his dark chocolate confections, while at night he experimented with different ratios of milk powder and cocoa powder. Friends told him it was impossible: the water in the milk would never mix with the fat in the cocoa bean. He could afford only one member of staff, apart from his wife. Using very basic water-powered machines, he tried every variation of evaporating the water from the milk before blending it with the chocolate. At one stage he thought he had succeeded. 'I was happy,' he said later, 'but a few weeks later as I examined the contents an odour of bad cheese or rancid butter came to my nose. I was desperate, but what was I to do?' Whether using milk or milk powder the result was a gritty, gravelly pulp, which had to be consumed quickly before it turned rancid. 'I did not lose courage,' Peter said, 'but continued to work as long as circumstances allowed.'

Meanwhile, Henri Nestlé's company continued to grow. By 1871 around a thousand yellow cans of infant milk rolled off his production line each day. Three years later this number had risen to half a million cans a year, that were being sold on five continents. With all the problems of manufacture, distribution, shipping and sales, managing the firm was becoming too demanding for Nestlé, who was approaching sixty. In 1875 he sold his company for one million Swiss francs to a Swiss businessman who bought everything, including the rights to Nestlé's name. One of the witnesses to the sale was Nestlé's friend and neighbour Daniel Peter.

Henri Nestlé understood the difficulties of managing the production of milk products in bulk, and advised his struggling friend to approach his own leading competitor, the Anglo Swiss Condensed Milk Company, which had found a way to mass-produce condensed milk. If Peter used condensed milk rather than whole milk, there would be less water to evaporate, and it might prove easier to remove the excess with the technology available to him.

When Peter met the managers of Anglo Swiss, it appears that he was reluctant to reveal much about his new idea. He was making 'a new product for which I am sure there will soon exist great demand', he said, adding that it was in their mutual interest for him to order the milk he needed from them. But when he repeated his experiments using condensed milk, the result, although improved, was still not reliable. He created a special 'drying room' where the milk-and-chocolate mixture was turned into flakes, spread on trays and heated further. Finally, in 1875, he hit upon a formula that produced a silky smooth chocolate. It was the world's first ready-made milk chocolate drink. He called it Chocolat au Lait Gala Peter.

The latter part of the nineteenth century was the heyday of English tourism in Switzerland. It was common practice for the wealthy to send their daughters to Swiss finishing schools, or to take a tour of the Alps for the spectacular scenery. One English grocer who made his way to Vevey was so struck by Daniel Peter's new product that he placed an order for one hundred pounds of milk chocolate to try back home. Vevey was also on the route of the Orient Express, and soon Peter's invention, in its copper-coloured wrapper, was making its way east.

Peter seized every opportunity to promote his new product. In 1878 he took it to the Universal Exhibition in Paris, organised to celebrate France's recovery after the Franco-Prussian War. The largest exhibition ever held, it attracted thirteen million visitors over the six months it was open, many of them from across the Channel. A world of wonders awaited inspection, including strange novelties like Alexander Graham Bell's telephone and Thomas Edison's phonograph. Among these feats of ingenuity Le Premier des Chocolats au Lait, the brand-new chocolate drink, gratifyingly received a Silver Award. Daniel Peter had arrived.

Better still, the English loved Peter's milk chocolate, allowing him to dream of a secure future: 'If a town of 8,000 inhabitants can consume more than 100 pounds in a year,' he calculated, 'then the six million inhabitants of London could easily consume 40 tons.' Full of enthusiasm, he set off to England to challenge the Quaker firms and their pure dark cocoas with his sweeter, milkier brand. And why stop there? It occurred to him that he could turn his novel chocolate drink into a solid form for eating. If the milk and chocolate could be fashioned into some kind of tablet or bar, it would be far sweeter and smoother than the slightly bitter dark chocolates on the market. Could there be a future, he wondered, for a *milk chocolate bar*?

Peter's breakthrough milk chocolate drink was swiftly followed by another Swiss technical advance. Sixty miles from Vevey in the capital city, Berne, not far from the cathedral, was a small watermill run by an aspiring chocolatier named Rodolphe Lindt.

Lindt had trained as a confectioner, and was keen to try his hand at creating chocolate. Reputedly he had a greater taste for the pleasures of life than for the hard disciplines of business. A contemporary photograph shows an elegant man, his dark suit neatly pressed, his fashionably upturned moustache treated with stiffeners – not the kind of man to devote his spare time to a rundown factory.

According to the probably apocryphal story, on one occasion he left for his weekend entertainment in such a hurry that he failed to stop his roller grinder, which went on pressing the cocoa beans for a full three days. When he returned to his factory on Monday morning, he found that the resulting mixture, far from being ruined, was silken and smooth. The beans had been pounded and churned to an exquisitely fine texture, resulting in an irresistible velvety chocolate full of subtle aromas. Whether Lindt's famous recipe really had its origins in such

Rodolphe Lindt, c.1900.

a happy accident is not known, but there is no doubt that he was soon experimenting with his unique method of preparing chocolate.

After consultations with his brother August, a pharmacist, he tried variations in temperature and timing to see which yielded the best results and enabled him to fold extra cocoa butter into the mix. Cocoa butter, unlike other fats, melts at body temperature. After several days, folding in as much cocoa butter as possible, Lindt found he had a rich, smooth chocolate. Encouraged by the possibilities, he developed a special machine that he called a conch, due to its shell-like shape. It was a wrought-iron trough, firmly embedded in a granite base, with sides that curved inwards to stop spills. A heavy granite roller attached to a steel arm passed repeatedly back and forth over the chocolate paste. As the chocolate slapped against the sides of the trough and over the granite roller that drew air into the mix, it became lighter, airier, finer and more liquid. Lindt found that after three days of 'conching', still more cocoa butter could be folded into the mix, making the chocolate silky smooth. He proudly called it his Chocolat Fondant.

It soon became clear that conching also facilitated the manufacturing process. Rather than creating a solid dough that was hard to press into the moulds, Lindt could simply *pour* his liquid chocolate into moulds to set. A dark chocolate bar was no longer something gritty and toffee-like, that just might break your teeth. The genteel young ladies in the nearby finishing schools at Berne and Neuchâtel were delighted to sample the results of his experiments, and the news spread fast: Rodolphe Lindt's Fondant could 'melt in the mouth'.

Surprisingly, Lindt was in no hurry to exploit his chocolate speciality. Without a sales team, he relied on word of mouth to promote his product, and local connoisseurs were eager to test this new sensory

Nineteenth-century chocolate conching machine.

delight. Across town in Berne, Jean Tobler was so impressed by Lindt's chocolate that he tried to create something similar for his own shop, the Confiserie Spéciale. It proved hard to fathom the key to Lindt's success, so Tobler approached him directly in the hopes of joining forces. Reports of Lindt's breakthrough reached other Swiss chocolate entrepreneurs, such as Philippe Suchard, who had opened sweet shops in Serrières, Lorrach and Neuchâtel. The Sprüngli family, owners of a successful chocolate business in Zürich, were amazed at the quality of chocolate produced by this novice, and all the Swiss firms were keen to crack his formula.

But Rodolphe Lindt was so determined to safeguard the secrets of his unique conching process that he installed his new technology in a separate building, with the machinery protected as though it were the Crown Jewels. The key that guarded it, was guarded itself. The fabled recipe that had apparently come into being through a careless accident was firmly locked in the mind of Rodolphe Lindt.

Switzerland was fast establishing a reputation as the land of chocolate. Both Rodolphe Lindt and Daniel Peter had created recipes that set them apart, recipes so mouthwatering that they could destroy rivals – not just in Switzerland, but across the Continent. And after several trips to England, Peter had no doubt that there was an appreciative foreign market waiting for his wonderful new milk chocolate invention – especially in Britain's booming industrial cities.

7

Machinery Creates Wealth
but Destroys Men

In Birmingham, George Cadbury, like his older brother Richard, was more focused on his ideals than on foreign competition. He loathed the fast-growing slums, the dark and ugly industrialisation that was spreading like a suppurating sore over the unspoiled countryside. Why, he asked, should progress and the 'triumph of machinery' lead to a reduction in the quality of life? 'Machinery,' he declared, 'creates wealth but destroys men.'

George and Richard saw the degrading living conditions of the slum dwellers as a cause of their continuing misery, driving them to drink to escape the squalor of their lives. They saw the very worst of it in Birmingham at first hand through their work for the Adult School movement. Before the Education Act of 1870 there was no compulsory elementary education, and the vast majority of adults in the slums were illiterate. The aim of the Adult Schools was to help them learn to read and write.

Every Sunday from the age of twenty, George rose early to ride to Severn Street, in one of the roughest districts of the city, to meet the other teachers before school started at 7 a.m. The dedication of his fellow teachers proved an inspiration. The Quaker and former Mayor of Birmingham, William White, taught every week at the Adult School until his death at the age of eighty. On one occasion, George learned that White, seventy-five at the time, had walked 'a mile and a half through untrodden drifts of snow up to two feet depth' to get to his class punctually.

The pupils were equally inspiring. They came from the slums, the drinking houses, even the prisons, united by a desire to improve their lot through literacy. George's class swelled to three hundred students, and he taught more than 4,000 over a period of fifty years. One woman rose during one of his lessons and declared that she had had 'more happiness in the one year since her husband joined the class than she had had in the twenty-nine years of her married life before'. This kind of touching experience fuelled George's conviction that the best way to improve a man's lot was to raise his ideals. But how, he asked, 'can a man cultivate ideals when his home is a slum and his only place of recreation is a public house?'

George and Richard knew that the adults in their classes were preyed upon by 'the sweater, the rack renter and the publican' – the unscrupulous employers of sweated labour, overcharging landlords, and managers of public houses who exploited people's weaknesses and helped to drive them to ruin. Their outspoken opposition to these groups made them enemies. 'It is unreasonable,' George declared at a meeting in Manchester, 'to expect a man to lead a healthy, holy life in a back street or a sunless slum.' Was it necessary, he continued, that the factory system should 'narrow the lives' of the workers to such an extent, 'belittling and oppressing' them with hideous conditions of life? But while it was clear that the factory system was responsible for the country's growing prosperity, what was the solution?

The idea of social welfare and reform was still in its infancy. Apart from the dreaded workhouses, there was little means of support for those who fell into poverty. Many Victorians managed to turn a blind eye to the full horror of poverty on their doorstep until writers like Charles Dickens forced them to look. Driven by his own experiences as a child, when he was forced to work in a blacking factory, his novels highlighted the message that poverty is the cause of misery. Dickens believed passionately that 'the reform of habitations of the people must precede all other reforms, and that without it all other reforms must fail'.

In a Cadbury photograph album from the period, after pages of neatly displayed sepia portraits of family members, there are images of influential thinkers of the day. Included in the collection, besides Dickens, Charles Darwin and Thomas Carlyle, is the author and artist John Ruskin. At a time when many believed that the poor were themselves to blame for their circumstances, Ruskin was one of the first to question the role of the economy in perpetuating deprivation. In a series of powerful essays, 'Unto This Last', which appeared in the *Cornhill Magazine* from December 1860, Ruskin was so critical of capitalist economics that the magazine was forced to stop publishing them.

Ruskin argued for an ethical approach to economic transactions, and said that with wealth comes a moral obligation. A profit, he wrote in words that seem prophetic to this day, is legitimate only if it can be achieved without harming the greater good of society. His beliefs that every labourer should have a wage on which he can live, that all children are entitled to an education, and that land should be used to benefit everyone and not just the wealthy, were considered subversive and outlandish.

Ruskin's political and economic thinking coincided with George and Richard Cadbury's all-embracing religious sensibilities. Their unshakeable faith meant that all issues shrank to nothing in comparison with the choice between 'living for things of the spirit, or for things that perish'. The idea of material success for its own sake was abhorrent. They were determined to use their growing business in a way that was compatible with 'enlarging the riches of human experience'.

Full of idealism, George and Richard began to discuss ways in which they might actually do something practical to test out their ideals of factory reform. Did a factory have to be located in a slum? How could they raise men's ideals and help them improve their lot? How could they assist women and children to break out of the cycle of poverty? In the late 1870s the brothers began to nurture an idea: they could move their factory outside the city of Birmingham, and create 'a factory in a garden', where there was space, trees grew and the air was pure. It would be a model factory, with 'perfect friendliness among all'. Through nature, through a garden city, they would lead their workers to the 'celestial city'. They would build a New Jerusalem in England's green and pleasant land.

Critics scoffed at the idea of applying altruism to a commercial enterprise in this way. It was a 'wild adventure' that highlighted the 'rashness and folly' of the Cadbury brothers, and would end in disaster. Factories belonged in towns, and the very idea that fresh air could be relevant to a business venture revealed that the brothers were 'no more than fanatics'. The Cadburys' stated wish for their staff to find work 'less irksome by environing them with pleasant and wholesome sights, sounds and conditions' was derided. Who ever heard of 'pleasant and wholesome sights' making money? These men of God would soon be bankrupt.

The brothers did not listen to their critics. Each Sunday, on their day off, Richard and George walked along the railway lines out of Birmingham, looking for the perfect site. They knew these roads of iron were essential to their new venture. In barely a generation, more than 10,000 miles of track had been laid across Britain, the roaring steam engines shrinking the distances between towns. At last, in the spring of 1878 they found what they were looking for, an unspoiled fifteen-acre plot set in a rural landscape about four miles south-west of Birmingham, nestled between the villages of Stirchley, Kings Norton and Selly Oak.

From Stirchley Street station they approached the land down a quiet country lane, Oak Lane (now Bournville Lane). It was bounded on the east by the Worcester and Birmingham Canal and a branch of the newly opened Midland Railway. To the north, flowing through a buttercup meadow was a trout stream, which gave the site its name, Bournbrook. As Richard and George viewed this country idyll in the full blossoming of spring, their ideas began to take tangible form. Ignoring their detractors, they deemed the situation 'unequalled'. Their factory, they said, would be 'replete with every adjunct requisite for carrying out a growing business'. Here in this rural haven they would build a factory on the part of the land nearest to the station; it would be a factory that considered the well-being of the workforce, rather than just exploiting it.

On 18 June 1878, the Cadbury brothers bought the land at auction. William Higgins, a crème tablet worker at Bridge Street, remembered the day he heard the 'incredible news' that they were moving to the country: 'Hope sprang up in the hearts of everyone.' The excitement was infectious. Both brothers now had young families, and they were equally thrilled at the prospect. George proudly brought his wife Mary and their young sons to the building site in the works' horse-drawn van. Five-year-old Edward took boyish glee in the great quantities of mud and the mountainous piles of bricks. These visits made a strong impression, and years later Edward could still fondly picture his father riding on horseback along the country lanes, their temporary rooms at a nearby farm, and his newborn brother, George Junior, sleeping through the fun in his pram.

Richard's two eldest sons too, Barrow and William, would later remember being taken to bare fields near Stirchley, where they were instructed to 'dig holes with our spades so that the subsoil could be inspected'. Eleven-year-old William, although he was far more excited to find a stream full of trout, proudly recalled that he broke the first ground for the beginning of the factory. Sixteen-year-old Barrow, who was studying mechanical drawing in Manchester, was keen to show that he could help his uncle George with preliminary sketches for the designs of the roasting room, grinding mills, saw mills, engine rooms, packing room and chocolate room. Grandfather John Cadbury, now approaching his eighties and looking not a little formidable in a black top hat and dark cloak, his muttonchop whiskers now completely white, made his way slowly around the site with his walking stick. It was satisfying for him to see his two sons embarking on such a promising venture, and his grandchildren playing in the fields.

Determined not to allow costs to escalate and sink them into debt, George and Richard hired and managed their own building teams, supported by a young local architect, George Gadd. A nearby Quaker firm, Tangyes Brothers of Smethwick, offered practical help in the form of their foreman bricklayer. Tangyes Brothers were also on hand to assist with the engineering design. But no one could help with the weather. The first brick was laid in January 1879, and building started in earnest in March. But constant, drenching rain reduced the clay site to a dangerous quagmire of mud. 'The horses that were moving the soil were half buried in clay,' according to an account in the *Bournville Works Magazine*. 'Two or three of them broke their legs struggling to get out of it.' The word went out that the brothers had come unstuck already. They would never finish on schedule. George, however, kitted out in special long boots, would not be stopped by the rain. He supervised the works from first light, and returned at nightfall after spending a full day at Bridge Street.

William Higgins was just one of the members of the Bridge Street staff who regularly trudged the four miles to watch the progress, whatever the weather. 'So eager were the Bridge Street hands to get out here and our journeys so frequent,' he said, 'that we could almost

The Bournville building site became a quagmire.

tell how many bricks were laid weekly.' It took two million bricks
before the new 'fairyland factory', as Richard's son Barrow called it,
was close to completion. In just six months the brothers were able to
close parts of the Bridge Street factory. The female staff members
took a seven-week break while they transferred machinery by canal
right to the heart of the Bournbrook estate. Perhaps with a shrewd
eye to future marketing, at the last minute some bright spark suggested
changing the name of the new site from Bournbrook to Bournville,
lending it a French flavour, at a time when the products of French
chocolatiers were so highly admired.

In September, Richard escorted a party of women staff around the
completed works. According to his daughter Helen, he bought their
train tickets from the centre of Birmingham to Stirchley Street station
'like a father of a family taking his children out for an excursion'. As
they drew near the site, everyone was 'in a state of happy flutter and
excitement', with Richard eagerly pointing out landmarks. The party
stepped down from the train, momentarily stunned by the enveloping
silence of the country. At last they could finally see what their new
life would be like.

They picked their way down the muddy country lane, and the

factory came into view. It was just one storey high, Richard explained, to avoid the need to carry goods upstairs, and stretched over three acres. He and George had personally supervised the layout of the whole site. The coach house, stables and smithy were already complete. Beyond the works, a large field was set aside for the men to play cricket and football. There was a garden for the women, complete with swings and seats, and plans for shady pathways and 'other contrivances for outdoor enjoyment'.

The first Bournville cottages were built around 1880.

To the west, work was under way on sixteen semi-detached cottages for key members of staff. The plots allocated for each one were spacious, with front and back gardens large enough for their owners to grow vegetables. Behind the houses an orchard was being planted with 150 apple, plum, pear and cherry trees. In the fields beyond, where the River Bourn widened naturally into a pool, the brothers had plans to create an open-air swimming bath for the men.

Inside, the new chocolate works were unlike anything the workers had ever seen before. Gone were the cramped conditions of Bridge Street, with its awkward passageways, dark corners and steamed-up windows. The factory in the field was a revelation: a temple to space and light and order. There would be no more heavy lifting and carrying: a series of tramways were laid throughout to move goods effortlessly from room to room. The factory was lit mainly by skylights in the roof, although there were no windows in the southern wall to prevent it from becoming too hot in the summer. With mounting excitement the staff explored a series of large, airy rooms, their footsteps echoing in the cavernous space.

The roasting room was already in operation, a modern marvel where nine large cylinders driven by steam power rotated the beans over a coke fire. In the spacious milling room, a wonderfully scented creamy chocolate fluid emerged from heated granite millstones lining one side of the room. The packing room had 'a most ingenious American appliance' which could weigh and fill 20,000 packets of Cocoa Essence a day. The box-making department, also mechanical, cut board into the required shape and glued the various parts together. Two machines could make 12,000 packets daily. With orderly precision the chocolate-making department created delectable little chocolate treats which proceeded in neat rows on long conveyer belts to the appropriate box.

Finally the wide-eyed staff entered what was called 'the general girls' room'. Cathedral-like in its proportions, this vast auditorium of pure white space was dedicated to packing the Fancy Boxes. Here the women would be faced with nothing more tedious than guiding the tempting chocolate mouthfuls into their boxes. Every care for the staff had been taken in shaping the building. Beyond the Fancy Box room were kitchens with the latest equipment, designed to provide meals in minutes. Even more thoughtfulness came in the form of warm changing rooms, should the weather prove inclement. Richard and George's wood-panelled offices were linked by a private corridor. They planted a rose garden outside, beyond which the view opened onto a wide rural horizon.

Fired with renewed enthusiasm, the brothers had begun promoting some of their staff. Just five years after having been taken in as an orphan, William Tallis was appointed Works Foreman, with an office of his own next to Richard and George. Although Tallis had almost no education, he had 'natural abilities', observed one member of staff, 'which enabled him to rise from the ranks and take the responsible position'. The versatile Tallis could turn his hand to anything; one minute tackling an engineering problem, the next driving the works' pair-horse van with goods to the station. He even delighted the owners' sons by teaching them how to fish for trout in Bournbrook stream, recalled Edward. Other newly promoted members of staff included seventeen-year-old Edward Thackray, who had been in the firm for just three years, and who George began teaching the responsibilities of buying cocoa at the London auctions.

It took some weeks to make the move, but on 27 September 1879 Richard told his father, 'I have cleared out all my furniture from the old spot; table, books and safe are all deposited in my new office.' Perhaps sensitive to the twenty hard years his father had devoted to the Bridge Street works which he was now closing for the last time, Richard spoke of 'old associations bringing back past memories'. But, he continued, 'the world and time move on and we must move with them'. Finishing on a more optimistic note, he added, 'The mess room was opened last night for the first time for the girls' tea and it answered admirably. About a hundred who stayed later than the others availed themselves of it.' Later the mess room was put to the test, delivering eighty chops in ten minutes.

But for all the enthusiasm the brothers and their staff felt about their new country retreat, nature found a way of making its presence felt. The early autumn brought a plague of wasps, despite concerted efforts to destroy all wasps' nests in the district. This was followed by a winter that was bitterly cold and wet, testing everyone's resolve and revealing unwelcome anomalies in the heating of the factory: some areas proved too hot, while others were icy cold. One employee recalled his surprise at seeing 'Mr Richard or Mr George go down on their knees crawling under the tables to see if the water pipes were hot enough'. The brothers' fatherly interest, he added, 'made a great impression on us'. On wet days, George used to check with the fore-women in each department that all the girls in their charge had changed into dry shoes. There was one unexpected benefit of the icy weather. The two ponds in the neighbouring estate of Bournbrook Hall froze over completely, and the Martin family who owned it permitted the Cadbury staff to skate there. 'Skating is associated in the minds of many with the first year here,' recalled one worker.

The damp and the cold were not the only difficulties. When orders

The works forewoman.

picked up before Christmas, George and Richard 'were at their wits' end to know how to execute them', said chocolate worker Fanny Price. Eventually they decided there was no alternative but to start the early shift at six in the morning. Although George had negotiated special rates with the railway company for workers' fares to Stirchley Street station, it refused to provide an early train. Many staff on the early shift had to walk from Birmingham across fields and muddy lanes in the dark – some rising as early as four in the morning to be sure to arrive at work on time.

As the critics had predicted, getting the workers home was also a problem. At the time the station was just an open platform with no shelter, and Fanny Price recalled a system the brothers adopted to prevent the women from getting wet: the staff would wait under a temporary shelter near the old station lodge, and 'Mr Richard used to blow a whistle to intimate that the train was coming.' For those who missed the last train to Birmingham, getting wet was among the least of their worries. Crème room worker Bertha Fackrell remembered 'rough times' that first winter, with the women walking through quagmires in the pitch black 'in twos and threes, arm in arm, groping our way along'. There were no street lamps for much of the route to Birmingham, so a foreman with a lantern sometimes escorted the women home. Once the works cottages were complete, the brothers set up makeshift sleeping arrangements with bedding and pillows for more than twenty girls, and rooms were also found in the surrounding villages. 'We were all very happy,' said Bertha, 'for everything that thoughtful kindness could do for our comfort was done.'

Despite his kindness, Richard occasionally had a hot temper, 'which was a natural part of his energetic nature', said his daughter Helen. It took the form of flaring up suddenly over minor irritations, for which he would always apologise soon afterwards. In the early days of Bournville, Richard took to wearing a jaunty little black cap. This became 'a well known weather vane through all the works as to whether things were going right or not'. If the cap was placed correctly on his head, 'everyone knew that things were going smoothly and well'. But if it was tucked under his arm, 'it was a sure sign that something was brewing'. Worst of all, said Helen, 'was if it was being screwed and twisted in its owner's hands'.

And there was much cause for the cap to be wrung during the

·early days of Bournville. Apart from the expected teething troubles of installing new machinery and managing the transition from Bridge Street, the sheer altruism of the brothers could itself create problems. Since they managed their own building programme, local slum builders had no chance to benefit from the large venture. Their decision to introduce a temperance zone around the works also cut out prospective publicans. But what really fuelled the ire of Birmingham's slum landlords and publicans were the brothers' continued efforts to help those in the inner-city slums through the Adult School. When George unexpectedly received a summons from a local policeman while on his way to the Adult School, many at Bournville believed the policeman was in league with local rack renters. George's offence was no more than to lead his horse onto an empty footpath because the frost had made the road treacherous. In due course the policeman was himself linked to serious crime and imprisoned, confirming the suspicions held by some.

George was keen to push through another huge gamble which was doubtless the cause of much cap-wringing on the part of his brother. By any standards it was a bold move. Although their adulterated cocoas were still making money, they cancelled some of their most popular lines, such as Homeopathic, Pearl and Breakfast. By doing this they were handing their rivals a huge advantage, but George was clear: he wanted the Cadbury name to stand for quality and purity. But after these strong-selling lines had been cancelled, would **they** receive enough orders to support their cavernous new factory? The sceptics saw it as another irresponsible step.

Then there was the issue of the factory itself. No amount of altruism and good intentions could get around the fact that there was nothing quite like Bournville in England, let alone any evidence that it would work. Cynics watched the experiment in quality, fresh air and wholesome living, waiting for any sign of failure. Most thought it was only a matter of time.

In Bristol, the Frys continued to prosper. Francis Fry and his brothers saw no need to emulate the radical innovation of their Birmingham rivals. As they outgrew their sprawling premises in Union Street, rather than gamble on the huge investment of a brand-new factory, they chose to expand piecemeal, acquiring outlying premises, often at some distance from the main factories. Little by little, they acquired some twenty-four separate buildings of varying suitability, and chocolate goods in varying stages of manufacture were taken by horse-drawn vans from site to site through the busy, narrow streets of Bristol.

Francis Fry's team also saw no need for a rush to respond to Cadbury's pure Cocoa Essence. Cadbury's invention had made it possible for the public to make a chocolate drink at home as easily as a cup of tea, but the Frys followed only at a leisurely pace. Two years elapsed before they installed van Houten presses, and when they finally launched their pure new product, Fry's Cocoa Extract, they failed to see any need to promote it heavily. It was only when it became clear that Cadbury's pure Cocoa Essence was outselling Fry brands which had been established over generations that Fry's board began to ponder how they might seize the initiative in the adulteration debate. At the time that the Cadburys were moving into Bournville and taking the bold step of ceasing production of all their adulterated cocoas to concentrate their efforts on one superior brand, the Frys were still making almost fifty different types of cocoa, all of which had to be transported to and from various different premises during their manufacture.

With the advantage of the docks at Bristol, however, the Frys did excel at overseas expansion. Their chocolate tins reached British troops fighting in the Crimean War during the mid-1850s, and the navy found that Fry's cocoa, made without the benefit of the de-fatting machine, fitted their requirements perfectly. It was easy to transport, filling and nutritious, a worthy accompaniment to the ship's biscuit. Fry already had long-established trading links with Ireland, and the company pioneered sales in Britain's fast-growing empire.

In 1867 the British North America Act established Canada as a dominion within the British Empire. The Act also made provision for the Intercolonial Railway in Canada, which would make a rail connection from the port of Halifax on the Atlantic coast to the St Lawrence River. From here goods could reach the vast interior of the North American continent via the grand highway of the Great Lakes, opening up previously unknown commercial territory. Francis Fry hired agents to investigate.

He soon learned that there was little chocolate production in Canada. With Canadian grocers willing to try Fry's products, the company began to ship goods from Bristol across the Atlantic. From Medicine Hat to Moose Jaw, Grand Falls to Niagara Falls, colourful yellow tins of Fry's Breakfast Cocoa waved the flag for English-manufactured chocolate goods. For British immigrants and loyalists they brought great comfort during the bitter Canadian winters.

To the Fry management team, the 'Fry Spirit' built on centuries of Quaker values remained all-important. Concern for their workers' welfare was paramount. Unlike the Cadburys at Bournville, the Frys had no space around their factories to provide recreation grounds, but their wages were generous compared to many local firms, and they organised choral and dramatic societies, libraries and night schools. Apart from his work in the community, Francis Fry felt his religious responsibilities keenly, and this took a more esoteric form than it did for his Quaker counterparts in Birmingham.

As the years passed, he dedicated himself more and more to his long-nurtured dream to bring a faithful version of the word of God to the world. His quest to track down the original 'Great Bible' produced in 1539 by Thomas Cranmer was complicated by the fact that Henry VIII had ordered 21,000 copies, and that 'copies bearing the same date differed from each other in various parts'. He tracked down 146 copies – sometimes 'as many as forty lay open on the table at one time', wrote his son Theodore. No one 'so critically examined them, or recorded his labours with such exactness'. At a time when Charles Darwin's ideas were considered dangerous and morally corrupting, Francis Fry's efforts were valued by like-minded men, who praised his 'noble service' and 'unrivalled collection of biblical treasures'.

Around the time that the Cadbury brothers moved to Bournville, both of Francis Fry's brothers died. In 1878 Francis retired as chairman, and adopted the age-old custom of nominating the oldest son from the next generation, in this case his nephew Joseph Storrs Fry II, as his successor. This Joseph was cut from the same cloth as his uncle Francis. Shy and introverted, he did not marry, but channelled his energy almost exclusively into two areas: the chocolate business and the Society of Friends.

As a child, Joseph Storrs Fry II had been mocked at school for his plain Quaker dress. This experience did not shake his adherence to the strict rules and values of the Society. Like his uncle, he was keen to keep Quaker tradition at the centre of the business. His daily routine was, in the words of a relative, 'extraordinarily conservative'. Rather than strike out in bold new directions, he dutifully followed his uncle on many issues.

So the chocolate factory continued to use various premises scattered across town, regardless of their suitability. The packing department moved to an old Baptist chapel, while numerous other departments jockeyed alongside each other for space. Although some new products were produced, there were no notable innovations. The Bible study and hymn singing continued. Joseph Storrs Fry II was a keen champion of causes such as the Bristol City Mission and the campaign for the suppression of the opium trade. According to Fry company records, 'his charitable gifts were almost numberless', and like his uncle he was an enthusiastic supporter of the British and Foreign Bible Society. Also like his uncle, he remained convinced that there was no need for change.

While the house of Fry sailed blithely through the swiftly changing markets of the late nineteenth century, and the Cadbury brothers embarked on a series of calculated risks, Joseph Rowntree in York struggled to keep his business afloat. In the 1870s his sales list included all manner of delicious temptations, including Shilling Cocoa, Shield Chocolate, chocolate drops, Half Penny Balls, and other more wholesome foodstuffs such as Food for Infants and medicinal fruit salts. Yet there was nothing that quite enraptured the northern palate.

Joseph Rowntree persisted in his resistance to innovations such as advertising and marketing. This extended to keeping a watchful eye on shop-owners who bought the right to rebrand Rowntree products. One such owner, a Mr Blanks, relabelled some Rowntree's Homeopathic Cocoa, and took the liberty of adding a few words to the package that might actually prompt customers to buy it. The luckless Blanks soon heard from his supplier. 'It is *not* a pure ground cocoa,' Joseph stormed. 'It is *not* produced from the finest Trinidad Nuts. It is *not* the "best for family use". In fact the whole thing is a sham, not very creditable to anyone concerned with it.'

Joseph Rowntree not only wanted his consumers to benefit from his high trading standards, he also wanted to bring benefits to his workers. He and his brother Henry had a reputation as fair employers, but with the business struggling in the depression of the 1870s, good intentions were not enough. Although Rowntree's total sales rose from £7,383 per year in 1870 to £30,890 in 1879, the average net profit was only £372 a year. The years 1873 and 1876, when the company suffered losses, were particularly bad. It is no wonder perhaps that in the mid-1870s Rowntree still saw himself as a 'Master Grocer' first, and a cocoa manufacturer second.

As a plain Quaker, Joseph Rowntree appreciated thrift and frugality. Combined with a lack of a clear vision, this might have driven the business into the ground were it not for a curious and unexpected visit. In 1879 a French confectioner named Claude Gaget arrived in York, and asked to see Joseph Rowntree. Gaget had worked for a Parisian sweet firm, Compagnie Française, and had adapted one of its recipes to create his own version of a new kind of sweet that was popular in France, but was not yet being made in England: a fruity, chewy pastille.

It is hard to imagine the middle-aged Joseph Rowntree, a Quaker who had proved particularly set in his ways, tasting this eager young Frenchman's pastilles. Perhaps it was the influence of his younger brother Henry that had led to Gaget receiving a favourable reception. But as Joseph curled his tongue appreciatively around this small, colourful, fruity sweet, unlike anything he had tasted before, he caught a glimpse of deliverance. This was manna from heaven, and a Frenchman was responsible. God surely worked in mysterious ways. Here was a creation that might just enable him to get one up on both the French *and* his Quaker rivals.

Joseph Rowntree was not a man to leap into anything suddenly in a big way. With a very careful eye on the budget, he supported the development of the fruit pastille in a sober and responsible manner. Investing in a boiling pot was much cheaper than buying van Houten's expensive machinery. There was room for a boiling pot or two in a corner of the factory, and money could be found to hire an assistant for Gaget. Joseph knew that the French had a monopoly on pastilles and gums. If Gaget succeeded in creating a recipe that would allow him to mass-produce an expensive French speciality, he could launch a quality product at a low price.

Yet after much labour over steaming cauldrons of fruit and sugar, the samples Gaget delivered in 1880 did not live up to his promises. The texture was not right. The flavour was less than perfect. Joseph spurned sample after sample. The quest for an original breakthrough product remained as elusive as ever.

Meanwhile, at Bournville, George and Richard suddenly found themselves in the fast lane. 'Orders after a time came in so fast,' observed chocolate maker Fanny Price, that the spacious new factory 'at length became overcrowded'. Any fears that the decision to stop manufacturing adulterated lines might seriously harm sales were soon put to rest. The message that Cadbury stood for quality, and produced pure cocoa in wholesome conditions, distinguished the brand clearly from the rest of the market, and boosted sales. The brothers advertised for staff in the local villages of Kings Norton, Stirchley, Northfield and Selly Oak, and within a year of leaving Bridge Street the number of employees jumped from 230 to more than three hundred.

A comparison of Fry's, Cadbury's and Rowntree's sales figures during this period shows just how wrong the critics had been to doubt the wisdom of the move to Bournville. In 1875, Fry's total sales were £236,075, while Cadbury's were much smaller at £70,396, and Rowntree's smaller still at £19,177. Five years later, in 1880, Francis Fry saw his business grow to sales of £266,285. What he could not know was how fast Cadbury was catching up. That same year, Cadbury had sales of £117,505 and Rowntree of £44,017. In the nineteen years since George and Richard had taken over from their father in 1861, they had turned a loss-making firm into one that was almost half the size of Europe's leading cocoa manufacturer.

Richard's oldest son, Barrow, wrote of the collaboration between his father and his uncle: 'No two partners ever worked in more complete harmony than the two Cadbury brothers, Richard and George.' Their material rewards had not meant that the brothers abandoned their dedication to long hours and Spartan self-denial. According to family records, it had long been their custom to make a small leg of mutton last for an entire week of their meals at the factory: roast on Monday, minced by Wednesday, and 'using the bones and any scrap end to furnish the meal on Friday'. When their father John came to visit Bournville one day, the mutton bones were bare. He gave his sons a gentle rebuke, pointing out that this was not acceptable fare for the young clerk who dined with them, so ending 'the tyranny of the leg of mutton'.

Four years after the move, a reporter from the *Midland Echo* made the trip out to see what the brothers had created. 'In the midst of green fields, with the ripple of the brook-like Bourn on whose banks the kingfisher and the moor hen find a home, Bournville forms the central part of a natural picture as refreshing to the senses as is the cup of cocoa manufactured there,' he enthused. He was impressed by the creepers and shrubs 'evidently glorying in the pure air', and by the 'well dressed happy looking girls trooping in at the door'. He noted that although the factory was free of smoke, the firm had made such effective use of 'mechanical appliances' that 'the cocoa or chocolate is hardly touched by the hand of an employee'.

Inside, everything was on a larger scale than Bridge Street. The reporter described the roasting room, its giant revolving cylinders 'pervaded by a fragrance such as one might expect to arise from thousands of plum puddings just turned out'. After the husks had been removed and the beans ground, the chocolate, which now had the consistency of cream, reached large mixers where it was combined with vanilla and sugar. He was evidently also struck by the quality of the staff. In the 'light and cheerful looking' box-making room he found 'ladies in white blouses with voices as gentle and feminine as those of any lady shopping in New Street'. Every one of them, he remarked, looked 'happy, intelligent, well cared for, neat and clean as if they were out visiting'. As for the brothers themselves, 'there is so little suggestive of the factory owner about them and so many implications of benevolence and kindly feeling that one becomes irresistibly impressed with the thought that money for themselves is the last thing on their minds'.

But business growth was indeed on their minds. News of the Swiss breakthroughs had prompted George and Richard to create a research department to develop new lines. They hired a Parisian chocolatier, Frederick Kinchelman, known to the staff as 'Frederick the Frenchman', to refine such delicacies as Nougats-Dragées, Pâté Duchesse and Avelines for the Fancy Boxes. They also decided to open a shop in

Paris, their Quaker scruples being relaxed sufficiently to acquire expensive, prestigious premises at 90, rue du Faubourg Saint-Honoré. It was time to show the French some real chocolate confections.

Richard and George were also thinking about expanding into the far-flung British Empire. Their first traveller to venture overseas was Simeon Hall, who had visited Dublin in 1873. Now, following Fry's lead they worked through firms of exporters to set up sales further afield. In Canada, Edward Lusher, a local agent in Montreal, was hired to promote their goods. This was followed by a similar deal in Chile with Brace Laidlow & Co. of Valparaiso, who were sent a small sample of goods, the labels duly translated into Spanish. In early 1881 the brothers took their foreign ambitions a step further. The Frys were not yet in Australia, which presented them with an opportunity to break new ground. Instead of hiring local agents they sent out one of their own staff. A solitary traveller, Thomas Elford Edwards, was dispatched to cover the whole of Australia and New Zealand. He was the firm's first permanent overseas representative. His mission: to find out whether the Australians had any interest at all in chocolate. In July 1881 a letter arrived at Bournville bearing an Australian stamp. It was from Edwards's office in Melbourne, and it gave details of his first order. For the brothers it was a triumph, and the first tentative threads of a chocolate empire reaching to the other side of the world.

Long before the concept of 'globalisation' had been thought of, Richard and George recognised that their firm stood at the threshold of something big. They wrote of an 'extraordinary food revolution' that would transform Western lives with the coming of 'internationality in food'. For a family that just a few generations earlier had exemplified Napoleon's 'nation of shopkeepers', the Cadbury brothers seemed to be about to transform their firm into a much larger enterprise. 'There is certainly untold pleasure in having to contend with overwhelming difficulties,' wrote George, 'and I sometimes pity those who have never had to go through it. Success is infinitely sweeter after struggle.'

8

Money Seems to Disappear
Like Magic

The success of Bournville did not go unnoticed by one impoverished entrepreneur from Pennsylvania, in America. In 1879, the year that Richard and George Cadbury opened their 'factory in a garden', things were not going well for twenty-two-year-old Milton Snavely Hershey, who was having his first taste of failure.

The child of a broken marriage – a penniless father intoxicated by the pursuit of the American dream, and a careworn mother who had long since tired of it – the young Milton straddled the gulf between his father's wild ambitions and his mother's strict Mennonite background. His shirtsleeves rolled high, his overalls stained, his shoes scuffed, his trousers worn, he laboured over scalding mixes and gas jets to create confections of boiled sugar at his proudly named 'Spring Garden Steam Confectionery Works' in Philadelphia.

'Money seems to disappear like magic,' his mother, Fanny Hershey, reported to her wealthy brothers. She and her unmarried sister, Martha ('Mattie') Snavely, dedicated themselves to Milton's business, working through the night wrapping the sweets, but after three hard years Milton's candy shop was floundering.

The venture had started in 1876, after Milton had moved to Philadelphia with a few dollars sewn into the lining his coat and a great deal of optimism. At first he had met with success, benefiting that year from the Philadelphia Centennial Exposition. But despite his fashionable cards promoting his 'Pure Confections by Steam', there were three hundred confectioners in Philadelphia, and competition was intense. Within a few years, Milton was forced into making humiliating appeals to his wealthy relatives on his mother's side. The

Snavely uncles were inclined to favour the young entrepreneur in the family, but just as his business was making progress, his estranged father, Henry Hershey, arrived in town.

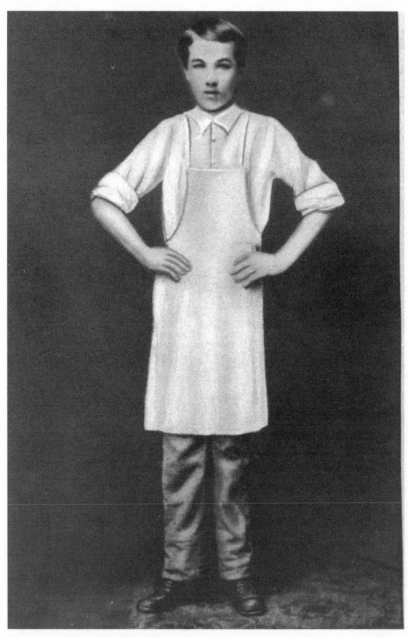

The young Milton Hershey as an apprentice at Royer's ice cream parlour, c.1873.

Forgiving and full of foreboding in equal measure, Fanny stood back and let her husband rekindle his relationship with his son. Man and boy became friends at once. Predictably enough, Henry, ever the dreamer, presented Milton with his latest money-making scheme. Improbably, it involved cough drops displayed in elegant little cabinets. Milton put every cent he had into this enterprise, and borrowed more, and then watched as Henry's scheme failed spectacularly.

Milton ended up giving his father several hundred dollars before Henry disappeared in disgrace to the silver mines of Colorado and another magical dream. Burdened with more debt, over the following months Milton and his mother were drifting towards failure. The pretty sugary sweets he made himself looked lovely in their rows of crystal jars, but he couldn't sell enough to cover his costs.

On 8 December 1880 he begged his Snavely uncles for another $600, explaining that otherwise he would not be able to pay his bills. His uncles obliged, only to receive another letter on 28 April 1881: 'I am in urgent need of $500.' On 3 December his Aunt Mattie wrote and asked for $400. This was followed a month later by a letter from Milton stating somewhat charmlessly, 'I must have $300 which Aunt Mattie says you are to raise and send not later than the early part of next week.' He said that against his aunt's advice he had given his father $350, which he now regretted: 'If only I had sent father on his way . . . as Aunt Mattie told me to.' The uncles paid up once again, but their good will was running out.

The humiliation of these appeals, the unrelenting hard work and the sense of obligation to his father combined to put intolerable pressure on the young entrepreneur. As his Philadelphia candy shop spiralled into decline during 1881, Milton Hershey too went into decline. Before long, he had become seriously ill.

❧❀☙

Milton Hershey had grown up trying to reconcile the inherent contradictions between his mother and his father; between Church and the hedonistic pursuit of wealth. His mother, Fanny, was the daughter of a bishop in the Reformed Mennonite Church, a faith that, like Quakerism, preached simplicity and plain living, although it differed in adhering to a literal interpretation of the Bible. The doctrine of hard work and self-discipline seemed crystallised in the very air his mother breathed, the plain clothes she wore and the strong back that bore the weight her iron will imposed. Her reward, she believed, would come after years of patient, honest work, and she never failed in her duty to toil on the path of righteousness and virtue. Sacrificing any temptation to gamble to small, purposeful steps was the creed that had helped her own family prosper. Her brothers had built up a comfortable lifestyle through their prudent endeavours.

His mother's background was in sympathy with the wider world in which Milton Hershey grew up. The state of Pennsylvania was Quaker country: plain, sensible and wholesome. It had thousands of Quaker inhabitants; indeed it owed its very existence to a Quaker, William Penn, who founded the colony in 1682.

During the seventeenth century, Quakers who had fled discrimination in England had faced equally appalling treatment in New England, at the hands of Puritan inhabitants anxious to keep them out. In 1659 the hostility erupted into violence: two Quakers who refused to leave, William Robinson and Marmaduke Stephenson, were marched to Boston Common and hanged. The following year, after two other brutal hangings, Charles II ordered the American authorities to stop their religious persecution. After this, Quakers were spared death, but just barely. There are reports of vicious punishments: ears sliced off, tongues pierced with hot irons, every brutality short of death.

The problem eased in 1682 when William Penn, the Quaker son of a distinguished English Admiral, set off for America. When Admiral Penn died, the British government owed him £15,000, but William agreed to waive the debt in return for a vast slice of land in America. He was given 45,000 square miles of rugged wilderness stretching from New Jersey in the east to the River Ohio in the west, from Maryland in the south to Lake Erie in the north. Penn called the new colony Pennsylvania, and welcomed not just Quaker settlers but other persecuted minorities from across Europe. On one of his trips around Europe he visited the Innesholden Alps – a German-speaking part of Switzerland – and persuaded the suffering Mennonites to come to the New World. Among them were Milton Hershey's forebears.

Penn's colony was founded on tolerance and religious freedom. The capital city, Philadelphia, the city of brotherly love, sat on the banks of the River Delaware and was a beacon for like-minded men. The

Quaker leader George Fox came to visit, and by the time of his death in England in 1691 there were 50,000 Quakers across North America, many of them in Pennsylvania – John Cadbury's older brother Joel would later put down roots in Philadelphia. Milton Hershey's great-grandfather, Isaac Hershey, established his Mennonite family in Dauphin County, near Derry Church.

Three generations later, the young Milton Hershey came under Quaker influence. With Henry Hershey's bad management the family moved constantly, eventually taking up residence on a small farm at Nine Points, near Lancaster, where Milton was sent to a Quaker school for a short while. The Quaker teacher reinforced the lessons of discipline, sobriety and hard work that Milton had learned from his mother. But many of the virtues espoused by both Quakers and Mennonites conflicted with a different and equally compelling message that the young Milton received from his father.

The Quaker schoolhouse that Milton Hershey attended.

Although Henry Hershey too had been brought up as a Mennonite, the iron discipline and self-denial that shaped his wife's character appear to have eluded him. He saw no need for the laborious incremental struggle that his wife believed was necessary to gain a modest foothold on success. There were short cuts, and he was out to find them. But Henry's schemes were not blessed with good fortune. His marriage, floundering for so long as a result of his failure as provider and the disappointments that followed one wild venture after another, finally failed. The trigger seems to have been the death of Milton's adored younger sister, four-year-old Serena. His parents, coldly distanced from each other and weighed down by grief, went their separate ways.

Henry Hershey could not give up his dreams. He needed them more than ever. Success, he told his son, came from imagination, from being bold, from taking on the world. The bigger the risk and the more grandiose the idea, the more astronomical the reward. 'If you want to make money,' he would say with easy confidence, 'you have got to do things in a big way.' And there was plenty of evidence to support his father's point of view. America was busy making money, with gold in the Klondike, Texas awash with oil and the wheels of industry ceaselessly turning. Now was the era of the dazzlingly wealthy: men like John D. Rockefeller and Andrew Carnegie, rich beyond imagining from oil and steel.

In the early 1880s, as Milton Hershey virtuously applied Quaker and Mennonite principles to his Philadelphia candy shop, the unprincipled John D. Rockefeller was holding the world in the palm of his hand. His oil kingdom was firmly rooted in wells across the north-east, the black gold that gushed from them daily a visible defiance of his humble origins. Rockefeller had started out in Cleveland, Ohio, sixty miles west of Pennsylvania, in the 1850s with nothing, and twenty years later was well on his way to becoming the world's richest man. This had not happened without adventure and risk-taking, or by the application of Mennonite principles. Rockefeller was the true-grit American for an aspiring businessman to copy. What young man could resist such entrepreneurial inspiration?

But in 1881, at the still struggling candy shop in Philadelphia, twenty-four-year-old Milton was battling to tackle his growing losses. The harder he tried, the more exhausted he became. He had given six mercilessly hard years to the business, and now, with failure imminent, it was taking him down too. His mother, always trusting in prayer, did what she could to help, but early in 1882 her sister returned from a trip to see her brothers to say they could offer no more money. By March, Milton Hershey had run out of funds.

His Snavely cousins brought their farm wagon to pack up the shop
and take Hershey back to Lancaster County. Disgraced and in debt
to his mother's family, he wrestled with his options. He was torn
between his father's self-belief and his mother's conviction that reward
came only through unremitting hard work. Once his health improved,
he turned his back on his mother's Puritan severity. Where had the
Bible, virtue and iron discipline got him? Milton opted for his father's
approach to business, and fixed his sights firmly on the west.

In the 1880s, while the Cadbury brothers at Bournville were finally
reaping the rewards of twenty years' hard work, their future American
rival was living hand to mouth among the wide landscapes of Colorado.
By the time Milton Hershey reached Denver, he was hungry and
desperate. When help from his father failed to materialise, he enquired
at a building in a backstreet that displayed a 'Boy Wanted' sign, and
was shown into a room where several dishevelled boys were waiting.
His instincts told him to leave, but he was locked in. Fearing that he
had been tricked into slave labour, he waited until the door was next
opened and drew his gun, which he threatened to use. He was learning
the hard way how to be tough.

During his travels west it is likely that he heard of an entrepreneur
in San Francisco, Domenico Ghirardelli, who had recovered from
bankruptcy by creating a business not too far removed from Hershey's
line of trade: chocolate. The vicissitudes of Ghirardelli's business
sounded all too familiar. He had arrived in California in 1849 during
the Gold Rush. When he failed to find gold, and his business selling
coffee and chocolate drinks to prospectors from a tent in the Sierra
foothills proved equally lacklustre, he went to San Francisco to open
a coffee and confectionery shop. This was wiped out in the great fire
of 1851, but the irrepressible Ghirardelli was soon back in business,
making a chocolate drink. He hit upon a low-tech process that enabled
him to achieve modest success in de-fatting the cocoa bean, which
gave him an edge over his rivals. The chocolate liquor was simply
hung in cloth, allowing the fatty butter to drip out while the cocoa
solids were left behind. Although this did not give Ghirardelli's
drinking chocolate the purity of van Houten's cocoa, it turned his

fortunes around. In the early 1860s he was importing just half a ton of cocoa beans a year; twenty years later he needed almost two hundred tons. His products were so popular they were selling around the Pacific, as far away as Japan and China.

For Milton Hershey, Ghirardelli's success on the West Coast echoed what he knew of the chocolate business on the East Coast. It was boom time – especially for New England's oldest chocolate-making firm, Walter Baker and Company. The business was started by Walter Baker's grandfather, a Dr James Baker who rented a mill in Dorchester, Massachusetts in 1765. Dr Baker's partner, John Hannon, was an Irish immigrant with experience in making cocoa, and they soon fitted out the mill with kettles, pestles and a large iron roaster. When the unfortunate Hannon perished at sea, the Baker family continued to expand the business. Their products mirrored their English counter-parts': cocoa was mixed with starch, arrowroot or sugar to mop up the fats, and they emphasised the medicinal value of several of their brands. Walter Baker, the third generation of the family to run the business, went to London and learned from market leaders such as Fry and the Taylor Brothers. In the 1840s he introduced a new line of chocolate sticks. 'I learned to make them in London,' he wrote. 'They are to be eaten raw or melted on the tongue to taste.' He pointed out that 'they are much more suitable for children than candy or Sugar Plums'. The firm's advertising also mirrored that of its English counterparts. In 1883 it introduced 'La Belle Chocolatière', a beautiful girl who decorated the lids of Baker's cocoas and chocolates in a range that had become known throughout New England.

Milton Hershey joined his father, but they had no money to rent a mill or invest in the expensive equipment they would need to start up a manufacturing business of their own. Their only asset was their broad experience. Henry had travelled widely, and had flirted with everything from painting to gold prospecting. His son's expertise was much more focused. Milton's mother had enrolled him at the age of fifteen as an apprentice in an ice-cream parlour in his home town of Lancaster, Pennsylvania. For four years he had learned how to trans-form boiling sugar and water into colourful temptations: lollipops, boiled sweets, fruit drops and many others. Six years of running his own business in Philadelphia had honed his skills; he had confidence in making all types of candy. Most recently he had taken a job at a sweet shop in Denver, where he refined his knowledge of caramels. Instead of adding paraffin to achieve the desired texture, his employer had hit on a better recipe. 'I put in fresh milk,' he said. 'They stay fresh for months and the milk makes them chewy too.' Milton was impressed, and took note of each step of the process.

The summer of 1882 found twenty-four-year-old Milton working in the stifling heat with his father in rented premises in State Street, Chicago. The city was booming: its population of five hundred fifty years earlier had exploded to half a million; meat packers and livestock owners, railroad and factory workers – all potential candy eaters. The world, it seemed, was coming to Chicago. Here a sharp-eyed entrepreneur could glimpse the new America. A network of railroads was turning Chicago into a key junction between east and west as well as north and south, with steamboats crossing the Great Lakes to connect by canal to the Mississippi River. With Henry's imagination and Milton's experience, the Hersheys could surely make headway. Their plan was to make caramels and cough drops.

There was just one drawback. It did not take Milton long to see what his mother had known all along: Henry was no team player or business partner. He was never there when he was needed, and could be relied upon for nothing. Before long father and son decided to go their separate ways. In the autumn of 1882, Milton returned east to his mother and aunt in Lancaster, Pennsylvania. He told them he wanted to start again. This time he would take on the largest growth market of all: New York. And he wanted to do it properly. As his mother had tried to teach him, slow, incremental steps were needed to grow a business. He knew that now.

Milton paced Manhattan's grid of streets researching the competition. America's largest city had everything: the immigrants pouring in from around the world formed a ceaseless tide of humanity, with hope in their faces and dust in their eyes. This was where the elusive American dream started, on grimy streets where wagons and horses mingled with dirt-poor Irish immigrants, New England Yankees, Germans and Scots from Pennsylvania, people from far and wide. The opportunities greeted you everywhere, blazoned across the billboards and in advertisements and eye-catching shop windows with a bewildering array of goods. There were new skyscrapers all of ten storeys high, as well as run-down tenements, with laundry strung between buildings soaking up the dirt. It was a wanting world, looking hard for fulfilment.

In the fashionable districts of downtown Manhattan, Milton found there was stiff competition. Countless candy shops already offered every kind of delight that could be whipped up from boiled sugar and water. Even novel forms of eating chocolate were on sale, such as chocolate drops, sticks and bon bons. Undaunted, Hershey leased a shop on Sixth Avenue, between 42nd and 43rd Streets near the elevated railway, and began labouring seven days a week in the basement kitchen. It was his second confectionery business, and this time it prospered. Feeling optimistic, he moved to larger premises on 42nd Street, only to find he had made a critical mistake: he had accidentally overstayed at the original shop by a few days. The landlord sued him for a year's rent. His mother and Aunt Mattie came to his aid yet again. Always willing to provide free labour, they wrapped and packed, determined to help him win through. But just when it seemed that Milton Hershey might finally turn the corner, his father turned up once more.

As before, the dashing Henry Hershey, full of enthralling dreams and an unreasonable amount of self-confidence, urged his son to seize the moment. Winter was coming. 'Flu was predicted. New Yorkers would need cough drops. Milton had to think big and bold, and take risks to match. Once again, despite all the Puritan homilies and all the hard work, he found himself unable to resist a gamble. He borrowed $10,000 to help him purchase the necessary equipment. Once again, the son's success and the mother's peace of mind were hostage to the errant father's dream.

⋙❀⋘

In 1882, the year that Milton Hershey embarked on his venture in New York, in Birmingham, Richard Cadbury's oldest son, nineteen-year-old Barrow, was due to start full time at Bournville. His uncle George suggested that he should first take a tour of the New World to investigate the chocolate market in America and Canada. 'It was one of Uncle George's generous and thoughtful propositions,' Barrow wrote in his diary. He was to travel with William Tallis, the works foreman, and young Barrow appreciated his companionship and business insights. Richard took them to Liverpool docks and saw them off on the beautiful steam-propelled White Star liner SS *Republic*. The ship had four masts, with full sails rigged on each one. 'I remember this', said Barrow, because it was a stormy crossing, and 'in a strong gale one of the top sails burst with a loud explosion'.

In chilly Montreal, where an ice railway had been laid for over a mile across the frozen St Lawrence River, William Tallis and Barrow Cadbury visited the company's Canadian agent, Edward Lusher. Montreal was in an excellent position to benefit from a growing network of railways and seaways that were creating paths through great swathes of virgin Canadian territory. To the west, cities like Toronto were already within reach, and the Canadian Pacific Railway was inching its iron feet towards the vast snowy peaks of the Rockies. Going east, rail and shipping routes made connections to the booming Atlantic coastal towns of Canada and the United States. But wherever Tallis and Cadbury ventured in this mighty landscape, they saw, cheekily stacked in shop windows, the bright yellow tins of Fry's Breakfast Cocoa.

Of all the cities on their tour, New York was the greatest wonder. Here Barrow Cadbury had a chance to see what Milton Hershey had patiently researched. Sugar candy was everywhere; chocolate was not. No one in America had yet tapped the full potential of the little black bean. If Barrow passed Hershey's dreary kitchen on Sixth Avenue, there was nothing yet to intimate a great confectionery rival in the making.

America, land of opportunities, lay in wait: a vast map studded with possibilities. Milton Hershey saw it, and believed that one day he would prove it. But as the young Barrow Cadbury, his horizons bounded by temperance movements and pacifist societies, picked his way through the brash American culture, he failed to spot the right opening.

9

Chocolate Empires

In the 1880s it was the mysterious continent of Africa that held Europeans in thrall. Unmapped, unknown, the opportunities it offered could be glimpsed from forays inland from coastal settlements. The land with the largest desert on earth also had an immeasurable tropical rainforest along the Congo, wide open savannahs burning in a shimmering haze and vibrant towns dotted along its eastern coast, set against the enticing blue of the Indian Ocean. The accounts of great adventurers such as Dr David Livingstone brought it vividly to life.

The British already controlled land along the coast of West Africa, including the Gambia, Sierra Leone and the Gold Coast. Now the empire builders of Britain saw the potential for a grand extension of Victoria's realm in a great swathe of land from the Cape of Good Hope in the south to Cairo in the north. But they faced rivals. Recently formed European states wanted to compete with established powers such as Britain, France, Spain and Portugal to seize their own colonies and build their own military and industrial might. Germany, formed in 1871, claimed land in south-west Africa. Belgium, created in 1831, eyed the great Congo Basin. Italy, founded in 1861, was manoeuvring for Abyssinia. The scramble for Africa was so intense that just thirty years later, only Liberia and Abyssinia would be independent of the Europeans.

The expansion of European empires in the late nineteenth century was mirrored by changes in trade – and chocolate was no exception. The exotic cocoa tree that had once thrived only in the Americas reached the shores of Africa. Portuguese colonialists were the first to take the hardier variety, *Forestero*, from Brazil in 1824 to plant on the island of São Tomé in the Gulf of Guinea. Spaced about a yard apart under the shade of banana or plantain trees, the cocoa trees flourished in the hot, humid climate to form a thick canopy. Cocoa eventually became the island's leading export, and *Forestero* spread to the neighbouring island of Príncipe, and made its way along the coast across the colonies of Portuguese Africa.

In Europe the price of cocoa beans began to fall as ever greater quantities were shipped from Africa and the Americas to satisfy the rapacious consumerism of the West. Delicious concoctions, once available only to the wealthy, were reaching an expanding industrial workforce through Dutch firms such as van Houten, and Peter and Lindt of Switzerland, in addition to the French and British manufacturers. In Britain, demand soared from just over a thousand tons a year in 1850 to almost 5,000 tons by 1880. And the English chocolate dynasties followed the trail of the colonial empire-builders as immense global horizons opened up.

George and Richard Cadbury found their new factory at Bournville was well placed to benefit spectacularly from the boom in global

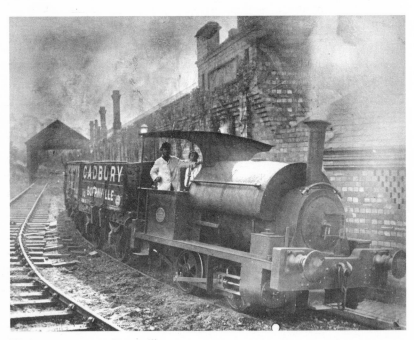

Works steam train at Bournville.

infrastructure. The railway at Bournville became double track and was part of the mainline into Birmingham, linking their chocolate works to all British ports. New waterways joined the canal route at Bournville, connecting it to both the Liverpool docks and the Bristol Channel, from which ever larger steamships sailed around the world. By 1880, Britain was linked to her colonies by almost 100,000 miles of cable spun under the world's oceans. Telegraph messages could be relayed across the world overnight. Cadbury, like Fry, began to explore links to the furthest reaches of the British Empire.

The Cadbury brothers' exploration of Africa began with a single traveller, Harry Gear. Gear had been pioneering sales in New York, but in 1886 he set sail for Cape Town. He soon wrote back asking for help, which arrived in due course in the form of a clerk from Scotland, one R.B. Brown, who answered the call to adventure. Brown's ambition was considerable, and he requested the whole of southern Africa as his 'patch'. Traversing wide tracts of land on horseback under a meltingly hot sun, he went where no confectionery salesman had ever been, carrying a stock of cocoa and chocolate wares that had never been seen before in Africa and blazing a trail from the Cape to Northern Rhodesia, Portuguese East Africa to South-West Africa, taking in Madagascar on the way.

The Frys also sent travellers to South Africa. Their literature shows they found the dangers of travel in Africa 'immense', and were stunned by the sheer scale of the unexplored interior. Nonetheless, 'the Dark Continent', they wrote, was beginning to open up to 'the enlightening influences of trade'. Their best markets were in the towns that sprung up around the diamond and gold mines. Fry's travellers found that the men working in the heat of the gold mines almost two miles below the earth's surface would start to shiver with cold once they were back above ground. They had a remedy for those chills: cocoa. Demand rose, and branches of the company were established in Cape Town and Johannesburg.

Meanwhile Cadbury's travellers were also making headway in Australia. Their first Australian traveller, Thomas Edwards, informed Richard and George that there was so much interest in their products that he required assistance. He was joined by William Cooper, a personal friend of George Cadbury. Their goal was wildly ambitious. The two travellers took a map and divided the continent between them. Edwards would promote Cadbury's goods in Victoria, South and Western Australia, Tasmania and New Zealand from his Melbourne office, while Cooper tackled New South Wales and Queensland.

Cooper launched his foray into the Australian market from a cramped apartment in a run-down part of Sydney. 'The houses or hovels are perched higgledy-piggledy on the rocks,' he wrote to his Birmingham colleagues, adding that from his window he had a splendid view of 'goats wandering about at will, chewing off the posters as high as they could reach'. It was an inauspicious start, but Cooper had a boyish sense of adventure and asked his brother, T.E. Cooper, to join him. Initial sales for the continent were little more than for a single small English town, but they grew rapidly.

Fry's overseas department followed Cadbury into South Australia. They promoted their products there in an original way, taking account of local rural sensibilities: 'Please shut the Gate and Drink Fry's cocoa' read the printed slogan they pinned to farm gates. Fry also ventured to India, where it faced difficulties arising from the huge variations in temperature from the tropical south to the much colder regions in the north. The same problem greeted the company's travellers in South America. Their cocoa drink proved popular at La Paz in Bolivia, which was at an altitude of 12,000 feet, but in the humid lowland plains they needed to devise sealed packages to keep the product fresh.

In the race to meet the global challenge, George and Richard Cadbury set up an export department in 1888, with a modest staff of six. Travellers spread to the far-flung corners of the globe. That year, their Australian traveller William Cooper enlarged his territory, acquainting the citizens of Ceylon, Calcutta and Karachi to the homely English fireside drink. His success encouraged the firm to assign a permanent traveller to the region, and they hired J.E. Davis to cover India, Burma and Ceylon. Mr Davis worked ceaselessly, in spite of illness, to bring the English drink to countries that must have seemed as strange and exotic as anything in the Arabian Nights.

But even Mr Davis would be outdone. Of all Cadbury's overseas travellers, Harold Waite from Birmingham was hard to beat. With magnificent determination and belief in his product, he won customers for the company on a grand scale as he trekked by train, by boat and on horseback through the West Indies and South America, before tackling the Middle East and then striking out east to Bangkok and Java. Unbelievably, it appeared that all those distant countries that the Victorians of Birmingham thought of as so bewitchingly mysterious were hungry for a taste of English cocoa and chocolate. Demand was so great that Richard and George Cadbury's travelling staff soon grew to fifty.

It was small wonder that the staff at Bournville looked forward to the annual travellers' party, held just before Christmas each year. The travellers brought back colourful tales of their great global odyssey: traversing fiery deserts and boundless distances, crossing alien continents carrying with antlike patience and determination the wares that brought with them a taste of European sophistication.

At the centre of all this activity, Bournville defied its critics. Cadbury's sales more than quadrupled in the 1880s, rising from £117,505 in 1880 to £515,371 in 1890. Richard and George seemed unlikely figureheads at the helm of this fledgling global enterprise. The manager of the General Office, Mr H.E. Johnson, affectionately remembered George 'with a row of small tins on a counter in front of him, the tins filled with roasted beans just bought in from the factory, and Mr George with unerring skill testing them and pronouncing judgment'. Sometimes Richard would join him during these tests of quality, and the brothers would sit together, happily absorbed in checking the batches. 'They consulted each other so much that no definite line seems to have existed,' continued Johnson. The two brothers were 'the centre round which everything moved. It was a kindly duocracy and those who served under it . . . have nothing but happy recollections of the early days of Bournville.'

Curiously, these eccentric philanthropists, who worked for the Lord and for whom abstinence and self-denial was a way of life, showed surprising flair in devising ever more enticing forms of chocolate indulgence. Apart from their drink, Cocoa Essence, their Fancy Boxes took increasingly luxurious forms. They were styled as a myriad of fashionable accessories: satin-lined jewellery boxes with elegant fastenings and locks, handkerchief cases or glove boxes, little cabinets to hold photographs or pens and other essentials. Each one was a work of art, with embroidered fabrics, painted illustrations and lace trims.

The rewards of running a growing business, however, presented fresh concerns for George and Richard that were increasingly in conflict with Quaker values. A Quaker business was meant to succeed, but to succeed quite so spectacularly was something the brothers had not envisaged. Creating personal wealth on an industrial scale was a problem. The turning wheels of industry at Bournville were making a fortune for their owners, but the Society of Friends' code of conduct had been fashioned in a different era, when manufacturing on such a vast scale could not have been foreseen. The austerity of the Quakers fitted better with a world where bounty for most people was still measured in terms of a good harvest.

Banking families such as the Gurneys and Barclays had gradually drifted away from the Society of Friends. They had come a long way from the days when the Scotsman Robert Barclay, a forebear of the banking family, was so inspired by George Fox that he wrote a defence of Quakerism, *An Apology for the True Christian Divinity*, published in Latin in 1676. When it was translated into English two years later it was acclaimed as 'one of the most impressive theological writings of the century'. Several generations later, the rise of mass consumerism brought unimaginable wealth to his descendants. It proved hard for them to reconcile the plain living of their Quaker ancestors with their growing wealth, and successive generations left the Quaker movement.

Apart from the issues raised by personal wealth-creation, the Cadbury brothers faced another challenge as their staff numbers continued to expand. The Quaker creed holds that everyone is equal: the 'inner light' is in each and every person. But how could the master and the worker continue to enjoy the close friendship they had shared in the early days of Bridge Street, when leisurely afternoons were given over to companionship, and employer and employee were indistinguishable. By the late 1880s, the firm had nearly a thousand staff. The sheer size of the business set the management apart, and worked against the formation of close ties with their workers. How did a Quaker firm acknowledge the inner light in each individual?

One solution to this had come from Joseph Storrs Fry II, who held a service for all of Fry's 2,000 staff each morning. In an exchange of letters with George, Joseph described the content and style of these services, including every last detail, such as the size of the hall, how to ventilate it, and the need for separate entrances for men and women. George and Richard found that the daily discussions they used to have with staff at Bridge Street evolved quite naturally into a service for their now larger workforce. One American visitor to Bournville, Dean Kitchen, was much moved. He described a women's service filled with 'a vast multitude all dressed in pure white and ready for a day's active service'. For him the 'short reading, kind words and simple prayer preceded by a hymn . . . was a revelation of religious purity and simplicity at full force'.

The service might set the tone for the day, but was it enough? George and Richard's father John, a plain Quaker to the last, embraced his beliefs so fully that even when he was old and in physical pain, he refused to exchange his hard, straight-backed wooden chair for an easier one, but submitted humbly to God's will. There would be no concession to personal comfort, whatever his need. He also insisted that his daughter Maria no longer dedicated her life to his well-being. Although it was too late for her to have children, she married in 1881 and left home. John meanwhile lived quietly near his thriving family. He was occasionally seen at Bournville with his stick, noting his sons' progress with pleasure while his own life remained a living testament to plain Quakerism and its insistence on the 'life of the spirit'.

⊶§✿§⊷

In York, Joseph Rowntree was not vexed by the problem of reconciling personal wealth with Quaker ideals. His business was still struggling.

The Frenchman Claude Gaget continued to toil over boiling cauldrons of fruit in the quest for the perfect fruit pastille, and Joseph and his younger brother Henry dedicated precious resources to finding just the right formula. Anxious words were exchanged as their early efforts were spurned, but by 1881 they believed they had cracked it; Gaget's recipe was luxuriously chewy and fruity. Advertising was unnecessary, the invincible Quaker Joseph told his staff. This was an honest product that he planned to sell at a fair price – a penny an ounce. How could it be other than marvellously successful?

But the launch of Crystallized Gum Pastilles in 1881 did not bring about an immediate change in the Rowntrees' fortune. Although demand rose steadily, the production costs kept pace. The Rowntree brothers had to order more boiling pots, and over two years the number of staff at the ramshackle factory at Tanners Moat doubled, reaching two hundred. If they dared entertain any hopes that they were at long last turning the corner, they were sadly mistaken.

The year 1883 was a difficult one for the Rowntree family. In May, after just a few days' illness, Henry died of complications from appendicitis. He and Joseph had run the business together for fifteen years, and Henry's cheerful presence had always balanced Joseph's seriousness. Now Joseph was alone with his worries at the helm of a business that had grown but continued to struggle.

Henry died owing money to the firm. His widow and three children needed some kind of modest support. And there was an outstanding sum of £10,000 in overdrafts and mortgages on the factory. Although sales jumped in the 1880s after the launch of the pastilles, Joseph fought to control costs. By 1883, the neat columns of red figures in his account ledger told a scarcely believable story. Sales had reached a record £55,547, but the company had still lost £329. Worse, Joseph knew the business was unstable. Sales for his cocoa, which he recognised was no match for Cadbury's Cocoa Essence and Dutch pure cocoa, could plummet at any time. Despite the unrelenting years of work, the future was insecure.

Joseph Rowntree had good reason to fear that his business could vanish altogether if he failed to develop a pure cocoa of his own. His first effort, which he called Elect Cocoa, had debuted without advertising in 1880. It was not a success, and was soon removed from sale. Fry had also launched a pure cocoa that failed, but in 1883 they relaunched it, and this time they made sure to get their message across. Fry's Pure Concentrated Cocoa won the backing of the *Lancet* and other medical journals, and was soon selling well. Joseph Rowntree, watching from York, could see that the market for pure cocoa was getting more crowded by the day and he had nothing with which to compete.

In 1885 Joseph embarked on a tour of Europe in an urgent quest to understand the Dutch process. Rising sales from his pastilles enabled him to invest in a van Houten press, but he did not yet know how to create a superior pure cocoa. In May he arrived in Cologne, Germany, to visit another Quaker firm, the Stollwerck brothers, and buy new equipment. Soon after, he was in Amsterdam, where he met Cornelius Hollander, who assured him that he had a process that could match the quality of van Houten's cocoas. The Dutchman was persuasive, and Joseph was unable to resist the promised gem of knowledge that might turn his company around. Deciding to take a risk, he agreed to pay Hollander £5 a week for six years; in return Hollander promised to 'communicate the secret of the van Houten manufacture of cocoa and make cocoa powder for Rowntrees'.

£5 per week was a handsome payment, and Joseph Rowntree did not want the rest of his staff to know how much he was paying Hollander. So when the Dutchman arrived in York and insisted on working in absolute secrecy, Joseph heartily agreed. At Hollander's insistence, his research room in North Street was carefully padlocked when he left each night.

Weeks turned to months, as Joseph Rowntree was obliged to watch and wait, while Hollander failed to deliver. Not only was no satisfactory recipe for pure cocoa forthcoming from behind his locked door, but his behaviour became increasingly strange. He guarded his workroom, overcharged for materials, burned his mixtures and exhausted everyone who dealt with him with endless haggling.

Finally, Joseph's patience ran out. He ordered some of his staff to break into Hollander's workroom, where they finally uncovered his secret: he knew next to nothing about cocoa production. With the police, Rowntree's staff then entered Hollander's house, and removed numerous objects which had been stolen or copied from the firm, including 'boiling glasses, drawings of hydraulic presses, drawings of the grinding mill, cocoa breaking machinery, a cocoa roasting machine' and more. It looked as if Joseph Rowntree had been taken for a ride.

It was not until 1887 that Rowntree relaunched a pure new Cocoa Elect. By now, four tons of pastilles were leaving Tanners Moat each week in horse-drawn vans, and demand showed no signs of flagging. When Joseph did his painstaking annual accounts he found that sales, at £96,916, had almost doubled since he had launched his gums and pastilles. After deducting costs, his profits were also rising – although at £1,600, they seemed a poor return for the huge volume of output. Had the company turned the corner? It was hard to tell. There were steps Joseph could take to help streamline the business, but at least some of them stood in opposition to his Quaker principles.

After Henry's death, Joseph was joined in the business by his oldest son, John Wilhelm. The young man could see the fragile state of the business, and was outspoken in his views. He pushed for change, his criticisms straying bluntly into areas that his father held most dear. There were aspects of Quaker thinking that were holding the business back, he told Joseph, and warned of 'Quaker caution and love of detail run to seed'. His arguments persuaded his father that changes had to be made, and a year after John Wilhelm joined the company, advertisements for Rowntree's products began to appear for the first time in popular magazines such as *Tit-bits* and *Answers*. This was just the beginning. As he worked his way around the firm's various departments, John Wilhelm blossomed into a natural deputy to his father. In 1888 he was joined by his younger brother, Benjamin Seebohm. Benjamin had read chemistry at Owen College in Manchester, and created a laboratory in which to experiment with new product lines.

Gradually Joseph's attitudes softened under the influence of his sons. He began to see modernisation through fresh eyes and to accept the need for change. It became clear that the business was being held back by the inefficiency of the raggle-taggle premises at Tanners Moat, which with their outdated machinery and many floors were a far cry from the gleaming, smooth operation at Bournville.

At first, Joseph prevaricated about borrowing money in order to move to a larger site. This was not Quaker philosophy as he interpreted it. Make do and mend. Thrift. That he understood. And as a Quaker, he saw another drawback to expanding the business. Joseph did not 'desire great wealth', he said, 'either for myself or for my children', fearing it could lead them into self-indulgence and greed. There was, however, one major consideration in favour of moving the business: he shared George and Richard Cadbury's view that it would be easier to improve conditions for his fast-growing staff at a site out of town.

In 1890 Joseph heard that there were twenty-nine acres of land for sale on the outskirts of York. Leaving the city walls and York Minster behind him, he followed a path for twenty minutes out of town, past rows of humble terraced houses, along Clarence Street, and into Haxby Road. He crossed a stream, and there on the left he found the site. Its potential was immediately clear. In these spacious grounds he could build the ideal chocolate factory, with plenty of room to grow. He could have a porter's lodge on Haxby Road, stables would be needed, and he envisaged tennis courts, a bowling green, parkland and lawns. With a special line built to the site by the North-Eastern Railway, the possibilities seemed endless.

Clearly it was a bold move that would require painstaking attention to financial details. But Joseph Rowntree now recognised that bold moves and investment were needed. At last Rowntree could match the progress of Cadbury at Bournville. After a long, hard struggle, Joseph and his sons were going to join the chocolate aristocracy. They would make pastilles, cocoa and chocolate for England.

While Joseph Rowntree had his eye on Bournville as a model for the way forward, the Frys continued to conduct their business as they had always done. Joseph Storrs Fry II felt the company's success was due to 'patience, prudence, honesty and hard work'. Worldly concerns were a distraction from the search for inner truth, and extravagance or self-indulgence could invoke God's displeasure. This guiding philosophy had served the family well for over a hundred years, and would carry them forward into the future. In 1885 Fry sold £404,189 of chocolate and cocoa. By 1890 that figure had nearly doubled to a staggering £761,969, and in five more years the company was approaching a million pounds in sales. It remained the undisputed, untouchable Quaker chocolate giant.

It is hardly surprising then that Joseph and others in Fry's management did not 'take readily to new fangled ways', according to the firm's *Bicentennial Issue*. The management believed there was no need to discard old methods that had been so spectacularly successful unless they were 'thoroughly assured that they had something better to put into their place'. Under Joseph Fry the company was doing well in areas such as overseas sales, but the innovative streak and initiative that had prompted his great-grandfather to create the business in the first place, and that drove his grandfather to pioneer the use of steam technology in cocoa production, were missing.

Furthermore, as a result of Joseph's desire to promote a Quakerly concern for all that was honest and true, Fry's advertising budget was less than that of Cadbury or Rowntree as a proportion of its sales figures. Fry favoured the gentler promotion of trade fairs rather than bombarding the consumer with advertising campaigns. The efficiency

Joseph Storrs Fry II.

of its production was also slipping behind the competition as the sprawling citadel around Union Street continued to spill out into any spare buildings, regardless of their suitability.

For Joseph Storrs Fry II the welfare of his workers remained a priority, despite their growing numbers. This extended to such thoughtful touches as giving each girl who left the firm to get married a copy of *Mrs Beeton's Book of Household Management*. He appreciated simple pleasures, such as the annual works outing, which gave the staff a rare opportunity to see a bit of the country. Long before the special excursion train was due to depart at 6 a.m., 'the platform would be crammed with Fry employees all dressed up in their best, many with flowers in their coats, anxious to set out on their journey to "furrin parts" ', according to one member of staff. As their numbers swelled, literally thousands of staff would descend on seaside resorts such as Weymouth, 'laying siege to all the restaurants' and almost reducing the towns to 'a state of famine'.

However, Joseph Storrs Fry II remained a plain Quaker at a time when many in the faith were wrestling with increasing conflict. The intellectual challenges posed by the writings of Charles Darwin left Quakers grappling with new clashes between religion and science, as well as enduring inconsistencies between personal wealth and religion. Recognising these growing dilemmas, the Society of Friends softened its rules. In 1860, for example, the Annual Meeting agreed that there was no need to follow 'plainness of speech, behaviour and apparel'. But Joseph valued his Quaker heritage too much to turn his back on it, even when his brother Edward gave up Quaker dress and custom to take a place at the bar. Edward proved so successful as a barrister that he also accepted a knighthood, a worldly accolade of the kind that Quakers traditionally spurned.

Little by little, almost imperceptibly, the paternalism represented by Joseph Storrs Fry II began to seem quaint, his religious values otherworldly, and the success of the business a miracle. Dressed soberly in his collarless black suit and waistcoat, a neat bow tie at his chin, Joseph appeared in increasingly striking contrast with those around him. Gradually, the head of the giant chocolate company, looking a little more tired and a little more grey with each passing year, seemed to belong to a world that was being left behind.

In the spring of 1887 the success at Bournville was marred by a deep personal loss for George Cadbury. On 23 March his thirty-eight-year-old wife Mary gave birth to her sixth child, a baby boy who died a few hours later. About a month later, the family was on holiday in Dawlish in Devon. One day George organised an outing for the children, leaving Mary behind to rest. He soon received a telegram saying that she had been taken seriously ill.

As was her custom, Mary made light of her illness, but a fever had taken hold almost unnoticed, and her condition deteriorated seriously. The local doctor informed George that there was nothing he could do. They must both prepare for the end, for that long separation, in the quiet, unfamiliar room, while the fever marked out those final hours on Mary's face. 'She was most patient during her illness,' George told the children later. 'She seemed willing to leave all in her Father's hands.' After twenty-four hours, 'her countenance was calm and peaceful as she passed into the presence of her King'. George and Mary had been married for fifteen happy years.

Richard's eldest daughter Jessie moved into George's house to help care for the five motherless children. In spite of her help, the loss of both wife and mother made for a sombre household. As the months passed, a distraught George turned to a friend from London, Elizabeth Taylor, confiding the terrible sense of loss that he and 'my precious little ones have sustained'. Elizabeth, or 'Elsie' as she was called by her family, had known George for over ten years; they had met by chance while she was visiting her uncle and aunt, George and Caroline Barrow, in Birmingham. Inevitably, it was Quaker interests that had drawn them together. George, who was organising a temperance meeting, had called upon the Barrows, and their young visitor offered to help out by making a speech.

Over the following years George and Elsie had met occasionally through the Barrows. She was inspired by his discussions of such thinkers as Ruskin and his practical vision of how social problems could be solved. He was impressed to find a forceful woman who was as passionate about Quaker values as he was himself. She taught a class of forty boys from the poor districts of south London on Sundays, in addition to organising choirs and Bible lessons, and had pursued her education rather than rush into marriage, taking over the role of governess to her younger brothers and sisters. Elsie was thirty when her father asked George to visit their home in London in the spring of 1888.

A whirlwind courtship followed. Elsie, with her even features, intelligent expression and high forehead, her brown hair swept neatly back, may not have been pretty, but was most certainly handsome. More important to George, she had a strength of purpose, an abundance of energy, and shared his passion for social reform. He felt confident that in this unusual and attractive woman he could build 'a life in the kingdom of the spirit'. They married at a Quaker meeting in Peckham in June 1888. 'The bride,' said the Ladies Pictorial, 'wore a gown of ivory satin trimmed with brocaded velvet, and a tulle bonnet with orange blossom and a veil. Her ornaments were a gold bracelet and a diamond and gold bracelet and brooch, the gifts of the bridegroom.'

It was soon clear that George had found a true soulmate. As soon as they returned from their honeymoon, in a clear signal of her intentions, she joined her husband at the Adult School in Severn Street. There, in the hopeful, anxious faces of the wives of George's students who came in to meet her, she could see her duty. The women asked if she could teach them, and Elsie happily agreed.

With her considerable experience teaching children, Elsie made a good stepmother to George's children. In March 1889 they welcomed a child of their own, Laurence. He was born just in time to be held by his grandfather John Cadbury, who died at the age of eighty-eight six weeks later. Another son, Norman, was born in 1890, followed in quick succession by Dolly in 1892, Egbert in 1893 and Molly in 1894. In 1892 George's oldest son by his first marriage, twenty-year-old Edward, had joined his father at Bournville, and was working his way up from the factory floor. Edward's cousins, Barrow and William, had already gained experience in the packing room, the chocolate room, the grinding room and the roasting room. To qualify for advancement, it was made clear to the younger generation that they had to understand the company 'in spirit' as well as the industrial and financial sides of the business.

In 1894, George's rapidly expanding household moved into a much larger home a couple of miles from Bournville. It was approached from the Bristol Road, where a lodge house marked the entrance to

George and Elsie with their family, c.1890. Back row: George Jr, Eleanor, Edward. Seated: Isabel, Henry, Elsie with baby Norman, George with Laurence.

a drive over a quarter of a mile long. Stately cedars and oaks bordered the track, affording intriguing glimpses of a grand house beyond. To the left was a large lake with an island, to the right, past a small copse on a gentle rise in the land, the rambling Victorian manor house suddenly came into full view. Nestled around the house were a series of gardens bordered with brick walls or herbaceous beds, including George's rose garden and a dairy. The scene encapsulated the rural serenity George loved.

This was not the home of a plain Quaker, but of a successful Victorian industrialist. The household and grounds staff numbered thirty, the women neatly turned out in starched white pinafores and caps. The former grocer's son, known to his friends as 'the practical mystic', had taken his place at the centre of his own little chocolate empire.

Among the Society of Friends there were purists who doubted that such an imposing residence could be reconciled with Quakerism. How did such luxury fit with the plain black coats, the frugality and Puritanical beliefs of the movement's seventeenth-century founders? Unlike Joseph Storrs Fry II in Bristol, the Cadburys did not adhere rigidly to the rules of their Quaker forefathers. But nor could they abandon their faith altogether, like some wealthy Quakers. Instead, they found a third way through the ever-widening gulf between the demands of their faith and the secular world. They belonged to a growing breed of successful Quakers who maintained their faith, but did not turn their backs on material prosperity.

For those in the Quaker community who considered the Cadburys too worldly, there was a surprise in store. In 1895 George was poised to use his wealth to pursue his ideal of building a utopia. It was the culmination of a family dream first discussed by his father and his uncle Benjamin nearly fifty years earlier. His aim was to turn his garden factory into a garden *city*.

Concerned that his plan might be thwarted if speculators and slum builders got hold of the land surrounding his factory, George had been discreetly using his income to buy parcels of land around the chocolate works. In 1893 he bought fields to the north of them, and then he purchased the imposing Bournbrook Hall, with its 118-acre estate, that adjoined Bournville to the west. With this land he was now ready to embark on the first stage of his ambitious scheme.

Initially he could only afford to build 142 homes around the chocolate factory. But it would be a start. He was convinced from his years working at the Adult School that if slum dwellers were provided with homes that gave them a sense of dignity they would thrive, and their health would improve. The key to his plan was land. Each home should have enough land around it for a family to cultivate a garden and grow food. This, he believed, would improve their quality of life and lead to a better diet. 'About a sixth of an acre is as much as a man working in a factory can cultivate in his leisure time,' he reasoned. Consequently, his village was designed with six or seven houses to the acre.

Poring over a map of the fields – Yellow Meadow, Far Hall Meadow, Barn Close, Fox Hill – George began to sketch out his plans. He appointed William Harvey, a young local architect, to help him. At

Bournville village green.

the heart of the model village would be a green, graced with established trees, winding paths and rose beds. The homes would be nestled around this green, each one individually designed to avoid ugly uniformity and set back from wide, tree-lined carriageways. Harvey was strongly influenced by the Arts and Crafts movement, inspired by John Ruskin and others to promote craftsmanship in decoration, furniture and architecture; William Morris was a member. The houses he designed for Bournville had a cosy English-cottage feel, with stepped gables, timber porches and Venetian windows over canted bays. Their spacious back gardens were 140 feet long, and planted with fruit trees to give a vista of blossoms in the spring. More than 10 per cent of the land was set aside for open spaces, including parks, lawns, tennis courts and playgrounds. On hand to help aspiring new home-owners who had never grown anything was a gardeners' department.

The model village served to complement changes at the rapidly expanding factory. In 1895 George and Richard turned land adjacent to the chocolate works into a men's recreation ground with a lodge. Soon plans were underway to build a cricket pavilion. It was on these grounds that the firm's cricket coach spotted the talent of Dick Lilley, an outstanding wicketkeeper who would go on to play thirty-five Test matches for England. On the other side of Bournville Lane, twenty-three acres around Bournbrook Hall were turned into women's grounds. The Martins' pond, on which the brothers and staff had fond memories of skating in Bournville's early years, was turned into a lawn surrounded by shady pathways, while a section was laid out for swings and other games. Plans were made for a swimming pool and an ornamental pond.

For George, the whole grand scheme was an experiment to demonstrate that rather than using land to benefit private individuals, it could benefit the whole community. Bournville village, however, was

Croquet in the women's recreation grounds at Bournville, 1896.

Ripening cocoa pods.

The Spanish introduced cocoa to Europe from the Americas. A chocolate pot can be seen on the left-hand side of Antonio de Pereda's *Still Life with an Ebony Chest* (1652).

The *Christian and Brotherly Advices* (1738) recorded decisions taken at meetings of the Society of Friends and set out codes of conduct for members. The section on Trading was the foundation of a written code of business ethics for Quakers.

In 1783 the *Advices* were updated and printed as the *Book of Extracts*. This page is from the 1801 *Extracts from the Minutes and Advices*.

The Quaker Frys of Bristol had created the largest cocoa factory in the world by the mid-nineteenth century.

Fry's cocoa was selling all over England in the mid-nineteenth century, at a time when the Cadbury brothers' business appeared to be failing.

"So near and yet So far."

The Cadburys' breakthrough product was a purer form of drinking cocoa, introduced in 1867.

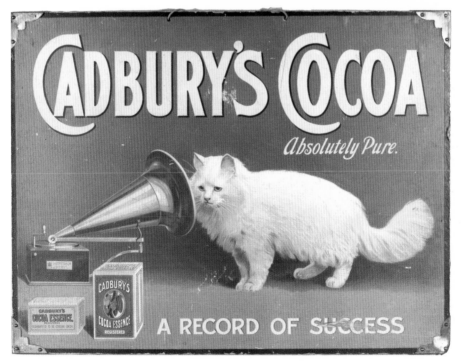

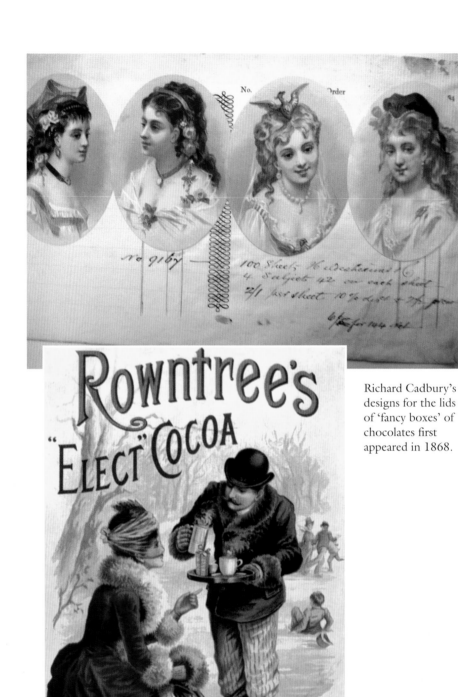

Richard Cadbury's designs for the lids of 'fancy boxes' of chocolates first appeared in 1868.

Joseph Rowntree brought out a pure form of cocoa in the 1880s.

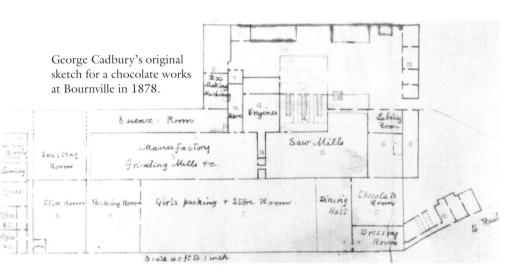

George Cadbury's original sketch for a chocolate works at Bournville in 1878.

Joseph Rowntree built the model village of New Earswick near York in the early twentieth century.

Tennis for female staff at Joseph Rowntree's
Haxby Road factory in 1900.

Staff swimming lessons at Bournville, c.1910.

Gala Peter
Le Premier
Chocolat au lait du monde

Hershey's
CHOCOLATE AND COCOA

In the late nineteenth century Daniel Peter in Switzerland became the first manufacturer to create milk chocolate for eating.

Milton Hershey, the American 'King of Caramel', staked his fortune on chocolate in the early twentieth century.

Rodolphe Lindt and Johann Sprüngli joined forces in 1899 to mass-produce chocolate from their new factory at Kilchberg, on the shores of Lake Zürich in Switzerland.

The American Forrest Mars made his first Mars Bar in Slough in 1933, and sold two million bars in a year.

not a charity. Houses were available at the cost price of up to £250. Loans were available at low rates: 2½ per cent if the applicant was borrowing less than half the value of the home, 3 per cent for a larger loan. The repayments for these loans were lower than the average rent, and enabled an employee to own his house outright after twelve years. In this way the would-be home-owner was not only encouraged to save, but could also aspire to a better lifestyle and a more secure future for himself and his family.

George Cadbury was not the first to try to create a model city. In 1853 a pioneering industrialist in the Yorkshire wool industry, Titus Salt, had constructed a model village at Saltaire just outside Bradford. In 1888 the leading soap manufacturer William Lever created Port Sunlight on fifty-six acres of marshy land on the banks of the River Mersey near Liverpool. Both schemes were designed to benefit their workers. George's plan, however, went further, in that he hoped to provide housing not only for Bournville staff, but for a broad social mix of people drawn from all walks of life. He wanted his model community to become a template for a means of raising the living standards of the poor elsewhere in England. Any investor, he believed, could create such a scheme without losing money, as he hoped to demonstrate by extending his model village using revenue from the estate. As building got under way in 1895, the homes proved so popular that George was soon negotiating to buy additional land.

But no utopia could survive without funds. The Cadbury brothers faced growing competition from abroad. The next challenge came from their old rival, the Dutch firm of van Houten, which had come up with a technical breakthrough that could threaten the success of Cadbury's chocolate empire.

10

I'll Stake Everything on Chocolate

At his factory in Weesp, Holland, Coenraad van Houten, who had sold George Cadbury his first cocoa press, had made another advance that had the potential to transform the cocoa business once again.

Van Houten wanted to improve the quality of drinking cocoa. He took a scientific approach to the problem, systematically testing out different ideas. It was known that the Aztecs had added wood ash to their preparations to counter the stringent acidity of cocoa. Working on the same principle, van Houten experimented with adding alkalis, such as potassium or sodium carbonate, during processing. The result took him by surprise.

When van Houten added alkaline salts before roasting the bean, he found the cocoa tasted less bitter. But to his delight there were other unexpected benefits in the texture and flavour of the drink. Although the alkalised cocoa was not completely soluble in milk or water, it was more so than any other cocoa product, blending more evenly in solution and becoming easier to swallow. Better still, the alkali enhanced the rich cocoa taste. The cocoa powder was darker, strongly aromatic and altogether smoother and more chocolatey. When he tested out his new drink, it was clear that the public loved it. He began selling it in the 1860s, and gradually word spread. No one knew the exact secret of the process, but it earned the nickname 'Dutching'.

By the late 1880s Cadbury's travellers were increasingly troubled by the growing presence of Dutch cocoa. Wherever they went, the smiling face on the packaging of van Houten's cocoa greeted them, proclaiming that it was 'best and goes furthest'. What is more, this domination of the market seemed to have been achieved effortlessly: grocers stocked Dutch cocoa simply because the public asked for it. For the Cadbury brothers, this could spell catastrophe. Since George and Richard had taken over the firm twenty-five years earlier, the brand that had done most to transform its fortunes was their drink, Cocoa Essence. But now their winning streak could be in jeopardy. Although sales were still growing as more people consumed cocoa, there was strong evidence to suggest that Cocoa Essence was achieving a smaller slice of a burgeoning market. By the early 1890s its market share was clearly in decline.

Richard and George were caught in a dilemma. They had staked their reputation, even their name, on the purity of their product. So how could they possibly start adding chemicals like alkaline salts? It would make nonsense of all their earlier claims. Worse still, van Houten's flyers showed that a distinguished line-up of British scientists had tested Dutch cocoa and claimed it was superior to anything else. Professor Attfield of the Pharmaceutical Society of Great Britain, Dr Theophilus Redwood, Emeritus Professor of Chemistry and Pharmacy, and Dr John Muter, former President of the Society of Public Analysts, were quite happy to sing its praises: 'The alkaline salt as introduced by the van Houten preparation, not only does not spoil but very greatly improves the cocoa both in its sensible and nourishing properties.' The eminent panel applauded van Houten's scientific approach, and concluded that his cocoa 'merits the term "soluble" more fairly than any other cocoa'.

George Cadbury, true to character, went on the offensive. Convinced that purity stood for quality, he set out to prove that alkaline salts were risky contaminants. The public must be alerted to the fact that Dutch cocoa had added chemicals. The first step was to print the message on Cadbury's packaging. An explicit warning appeared beneath the sweetly smiling Cadbury's girl assuring buyers that while this cocoa was pure, 'among the Cocoas that do not answer to this description are those of foreign make, notably the Dutch, in which alkalis and other injurious colouring matter are introduced'. Richard and George soon found experts willing to fight on their side of the battle, including members of the ever-obliging medical profession.

The *Birmingham Medical Review* of October 1890 was in no doubt where it stood. 'Quite apart from any question as to the *injury* resulting to the human system from taking these [alkaline] salts,' it stormed, 'it would only be right that the medical profession should resolutely discountenance the use of any and all secret preparations.' Scientists writing in *Peterson's Magazine* in 1891 went so far as to specify what complaints alkalis might cause: 'They dissolve animal textures . . . and excite catarrh of the stomach and intestines.' Dr A.J.H. Crespi went further, arguing that foreign cocoas with alkaline salts were 'dangerous and objectionable'. Even on the Continent – though not in Holland – Dutch cocoa was given the thumbs down. One anonymous German expert declared that Dutch alkaline cocoa was 'in the highest degree destructive', damaging 'the essential constituents of cocoa'. In short, he fumed, the Dutch method was 'perfectly barbarous'. The medical profession soon succeeded in making Dutch cocoa look like something that Lucrezia Borgia might administer to a rival.

But nothing could entice the public away from its favourite new chocolate drink. Sales for Dutch cocoa soared, reaching 50 per cent of the British market in the 1890s. George and Richard had to accept that purity was less of an issue than it had been in the past, when people had to worry about manufacturers adding red lead or brick dust to their cocoa. Now the public felt confident that their cocoa would not do them harm, they wanted a more enjoyable drink. The Cadburys had nothing to match what the Dutch offered.

And then the travellers arrived with more bad news. It came in the form of Swiss chocolate.

At Vevey in Switzerland, things had not been easy for Daniel Peter. After the initial excitement of discovering his revolutionary milk chocolate drink in 1875, it took almost twenty years to make headway in expanding his chocolate enterprise. His efforts to turn milk chocolate into a bar for eating were equally troubled. A key stumbling block was funding. In the 1870s he paid several visits to England, where he confirmed that there was a hearty appetite for his creamy, milk chocolate drink. Even so, he struggled to convince potential backers in Switzerland that his was a good business proposition. His manufacturing process was fraught with pitfalls that steely-eyed Swiss financiers could spot a mile away.

Just how hard it was to create a standardised milk-based product for export in bulk was all too evident. Milk was a tricky commodity to deal with. Thundery summer weather could turn it sour, and it often went bad before it could be processed. Its quality could vary from farm to farm and season to season. Attempts to manufacture a milk chocolate bar were equally problematic. Early efforts were dry and crumbly, and all too often the milk went bad, making the bar rancid. Convinced he had an excellent product, Peter searched long and hard for a financial backer, but without success. His friend Henri Nestlé had retired, and the new directors of Nestlé would not support him. Nor was he able to form a partnership with Anglo Swiss, the firm that provided him with condensed milk. Peter found himself in a financial wilderness, his product and its ingredients deemed too risky for Swiss bankers.

After years of low-budget experiments he finally mastered a technical process in 1886 that produced a temptingly soft and creamy milk chocolate bar. It was launched under the name Gala Peter, and received immediate acclaim. When demand far exceeded supply, bankers finally began to pay attention. Everything came together for Peter in the early 1890s when two Swiss businessmen, Albert Cuenod and L. Rapin, and a banker, Gabriel Montet, invested enough for him to create a new company and increase production: Société des Chocolats au Lait Peter.

'I think I can say with a pretty high degree of certainty that the majority, if not all, of the Swiss chocolate-makers have tried to copy me,' Peter proudly told his new board. 'All have had to give up!' The only rival product still in the running was a treacly milk chocolate paste manufactured by the Anglo Swiss Condensed Milk Company, but it was no match for the quality of Peter's goods.

With Peter's new resources dedicated to stepping up production and advertising, orders rushed in from across Europe. In the first six months of trading in 1895, sales doubled to ten tons of chocolate. The business was such a success that Peter and his team decided to recapitalise the company at a million Swiss francs, and they opened a second factory, which doubled their production capacity. The milk chocolate that had been a novelty luxury for wealthy travellers fifteen years earlier was becoming widely available. And no export market had a sweeter tooth than Britain.

British grocers took to Swiss chocolate as they had taken to Dutch cocoa. They could not get enough of it. The English Quaker chocolate-makers could not fathom how the product was made, and those who had wrestled long to make this 'food of the gods' had no intention of disclosing the recipe.

◆◦§✿◦◦◆

Daniel Peter was not the only Swiss chocolatier whose secret was becoming legendary. When Rodolphe Lindt built his unique conching plant at his factory in Berne, he had ensured that only a few of his workers had the key to the door. As the years passed, with no one able to surpass the quality of his chocolate, the mystique and intrigue surrounding his special process caught the public's imagination. A German magazine, *Gordian*, published an article in 1899 inviting readers to guess Lindt's special recipe. The magazine was inundated with letters. Did he have a new type of grinding machine to crush his beans to a finer texture? Did he beat his chocolate mixture for longer? Could it be the addition of essential oils like peppermint? No one knew. *Gordian*'s editorial team pronounced their verdict: Lindt's secret would never be known.

But Rodolphe Lindt, the gentleman entrepreneur, was in a selling mood. His business partnership with another Berne confectioner, Jean Tobler, had fallen through. Tobler despaired not just of Lindt's secrecy, but also of his company's erratic and small-scale production. Lindt, now approaching fifty, was flattered to receive offers from such competitors as the Stollwercks of Germany, who offered as much as three million marks, but in 1899 he opened the door of his conching plant to another Swiss manufacturer: Johann Rudolf Sprüngli of Zürich. Sprüngli had made a canny offer. Lindt would receive 1.5 million Swiss francs – worth roughly £60 million today – and be a director in their new joint venture

Johann Rudolf Sprüngli, described in company literature as a shy man, was anything but reticent when it came to business decisions. Shortly after inheriting his share of the Sprüngli chocolate business from his father, he had initiated a rapid expansion programme that culminated in 1898 in his moving the family firm from its cramped headquarters in the old town of Zürich to a new, modern factory on the shores of Lake Zürich by the railway at Kilchberg. The following year he joined forces with Lindt. Their new company, Chocoladefabriken Lindt und Sprüngli, proved to be a force to be reckoned with in the chocolate industry. For Rodolphe Lindt, it was a far cry from his inauspicious start in fire-damaged buildings in Berne. Together they would produce of some of the most acclaimed chocolate in the world.

In Britain during the 1890s, cocoa changed from being a product that only a few could afford to something that was on every household shopping list. Cocoa consumption more than doubled over the decade, from twenty million to forty-three million pounds a year. But the British chocolatiers feared that the Continental firms, with the miracle of chocolate fondant and milk chocolate, were poised to dominate their market.

The Quaker firms had established a clear lead over their other English rivals. Former competitors, such as the Taylor Brothers and Dunn and Hewett of London, had seen their profits slide. There were newcomers, especially in confectionery. John Mackintosh of Halifax postponed his honeymoon in order to save money to launch his toffee and confectionery business in 1890. Starting with a loan of £50, the 'Toffee King's' company was worth £15,000 in ten years, and was successful enough for him to build his red-brick Chocolate Works in Queen's Road, Halifax. Terry of York also prospered at its Clementhorpe works on the River Ouse, producing a considerable range of chocolates and sweets alongside its traditional candied fruits and peels. But during the 1890s, Fry, Cadbury and Rowntree were the dominant players in cocoa and chocolate.

By 1895, Fry had sales of £932,292. The Cadburys were close behind, with £706,191, and now there was heady talk: how long would it be before they caught up with Fry, or even overtook it? Both firms were among Britain's largest employers: Cadbury had 2,600 staff, and Fry over 4,000.

After a prolonged struggle, the Rowntrees were at last making headway. Their sales, at £190,328 in 1895, had more than tripled

The dining room at Bournville.

over ten years, and they were narrowing the gap between themselves and the two leading Quaker chocolate firms. The prodigious appetite for pastilles continued to grow, and was complemented by the successful launch in 1893 of Rowntree's clear fruit gums. But Joseph Rowntree knew his Cocoa Elect was struggling next to established brands of pure cocoa, and he had a huge investment in the site at Haxby Road. As he made the transition from a family firm into a large-scale manufacturer, his personal notes reveal that he watched the foreign competition anxiously.

To take on the Europeans, in 1895 Joseph Rowntree took the unusual step of approaching Joseph Storrs Fry II and the Cadbury brothers to discuss some form of collaboration. The companies had much in common, and were soon discussing policy on a number of issues. For example, at the time, shopkeepers could charge what they liked for a product, sometimes overpricing chocolate to increase their profits, or underpricing to undercut a competitor. The Quaker firms wanted shops across the country to sell their products at the price that was printed on the label: if it said sixpence on the packet, shop-keepers had to sell it at sixpence. By uniting on such issues, the three firms aimed to guarantee that distributors received good margins, but also that they could not go above them, to the detriment of sales.

Their informal discussions on discounts and shop displays ensured that a price war or a margin war did not break out between the English Quaker firms. They also hoped that this would help them to fend off the European giants. But Dutch and Swiss sales were strong, and the fast-growing Nestlé company was waiting in the wings: the battle lines were being drawn for Europe's chocolate war, with the unsuspecting British consumer the prize. The winners would be those who could devise the most irresistible confections to woo and win the English palate.

But as the European chocolate firms lined up to do battle, a newcomer on a different continent appeared with a plan for an enterprise that could dwarf the contest in Europe.

In 1893 Chicago, Illinois, was host to a great exhibition, the World's Fair. Twenty-seven million visitors flocked to view the most exciting inventions of the industrial world: engine-powered vehicles, electric lights, telephones, new appliances of every description. Among the crowds, one man kept returning to the machinery building: Milton Snavely Hershey.

There was one stall in particular that caught his eye. A German manufacturer, J.M. Lehmann of Dresden, had shipped his latest designs for making chocolate to America. Lehmann had a long-standing reputation for the quality of his machinery – in fact it was his firm that provided the engineering knowhow that helped van Houten develop his first cocoa press, which had revolutionised the cocoa drink. Now Lehmann had something that intrigued Hershey: a chocolate factory in miniature. The raw beans were roasted and then crushed by granite rollers, producing an aromatic stream of chocolate liquor. Sugar, cocoa butter and flavourings were added before the mix was set in moulds. Hershey was mesmerised, taking in every step of the process. After pondering it deeply, he turned to his cousin, Frank Snavely. 'Frank,' he declared, 'I'm going to make chocolate!'

Hershey knew that American cocoa imports, although modest compared to those of Europe, were rising fast. Europeans were consuming a hundred million pounds of chocolate per year, while Americans consumed only twenty-five million – but that amount that had almost tripled in ten years. Yet in just a generation, America had become the greatest industrial power in the world, outproducing Germany, France and Britain combined. The vast market of America would surely soon wake up and smell the chocolate.

Hershey was a gambling man, and by the 1890s he felt he was on a winning streak. The metamorphosis had happened slowly. After fourteen years of unending labour, two failed businesses and a lawsuit, things were finally going his way. He had come a long way from the turmoil of 1886 in New York. After borrowing $10,000 to help fund the sale of the cough drops his father so believed in, he found he was unable to repay the loan. Every day as he had toiled in his basement by the elevated railway, the gulf between his takings and his escalating debt widened.

According to a tale in the Hershey archives, in a last determined bid to raise money in New York, Milton hired a horse and wagon. With the cart brimming over with candy and cough drops, he went out in search of new customers. When he entered a candy shop, a gang of youths looking for some fun seized the moment to put fire-crackers under the horse. Milton emerged to see the terrified animal bolting up the street, the wildly veering wagon scattering its precious load. Having lost hope that he would meet the deadline to make one of the payments on his $10,000 loan, he sold everything and in the summer of 1886 went home to Pennsylvania. In a final coda to his experiences in New York, his carpet bag was stolen by a thief, and a friend had to pay his shipping costs.

Despite this second failure, Hershey had learned so much that he was determined to try again. Unfortunately, he found that his resolve was matched by that of his mother's family. No more money was forthcoming. The charming youth who had carried all their hopes ten years before had been transformed into a prematurely greying, twenty-nine-year-old disappointment. Money was not available. Not so much as a cent.

In the autumn of 1886, with no capital available, Milton Hershey had to start at the beginning again. He believed in his new recipe for caramels. It bore a remarkable similarity to the one he had learned in Denver, using milk instead of paraffin. No one on the East Coast was making a caramel quite like it. Hershey started once more, peddling his sweet dream from a pushcart around the streets of Lancaster, Pennsylvania.

This time, people came back for more. He couldn't make his caramels fast enough, and soon he could afford to rent a work space in
a warehouse. He had to share it with some noisy neighbours – a
carriage works and a piano manufacturer, amongst others – but he
had room for his kettles and sugar stores. Once again his mother and
his Aunt Mattie came to help with the sweet-wrapping. As word
spread, in a stroke of extraordinary luck, a British traveller passing
through town in 1887 placed a big order for caramels to ship to
England. Hershey duly set off to the banks in Lancaster to secure a
loan to buy the equipment that would enable him to meet the
deadline.

At first, no one would back him. But Hershey's luck finally turned
when Frank Brenneman, the cashier at Lancaster National Bank,
agreed to loan him $700. His luck turned again when he was unable
to repay the money within the agreed ninety days. Worse still, he
required an extra $1,000 if he was to fulfil the English order. The
records show that Brenneman, knowing full well that his superiors at
the bank would turn down this request, authorised the additional
funds personally, enabling Hershey to manufacture the caramel and
ship it to England.

Unknown to the bank, Hershey now owed it $1,700, and some
sort of payment was soon due if another slide down the ladder of his
life was to be avoided. As the deadline loomed, he knew he could
not repay the loan. The familiar sickening situation had caught up
with him again. Just as he was at his lowest point, a letter from
England arrived. No doubt it would be more bad news: perhaps the
ship had sunk, or the caramel was inedible. In trepidation he opened
it, to find a cheque for £500. He was still in business.

From that moment on, Hershey's Lancaster Caramel Company
started to reach out to the fast-growing industrial cities of the East
Coast. Somehow his family managed to keep his father at bay before
he could spoil it. Milton Hershey was soon in a position to open what
he loved to call his 'western branch' in Chicago, followed by another
branch in Pennsylvania, at Reading, and a shop on Canal Street in
New York City. By the early 1890s he was employing some one thousand staff. The surge in immigration, the explosion in consumer
demand and the skilful use of new types of machinery helped his
business to grow at an astonishing rate. 'Milton Hershey has made a

complete success of his life so far,' declared the *Portrait and Biographical Record of Lancaster County* in 1894. Noting that his company did over a million dollars of business a year with exports to Europe, Asia and Australia, it concluded: 'No man stands higher in business and social circles in the city of Lancaster than this man, who has been crowned with success.' It was quite a turnaround.

When Milton Hershey travelled to the Chicago exhibition in 1893, it was a very different experience from his earlier visit to the city with his father. This time he was a man of substance, with money to spend. A man who no longer lived from hand to mouth, but had a grand brick house in his home town of Lancaster filled with soft furnishings, fine art and other luxuries he had been denied as a child. On his last day at the exhibition he bought Lehmann's entire display: the chocolate factory in miniature.

In 1894, when George Cadbury was moving into his manor house, Lehmann's incredible chocolate machine was installed unobtrusively in the factory of the Lancaster Caramel Company, and started to produce Hershey's own chocolate. The first attempts were not promising: the results were coarse and dry compared to the Swiss milk chocolate. But Hershey was a great experimenter, and he began to test out different ideas: chocolate wrapped around caramel, chocolate cigars with exotic-sounding names like 'Hero of Manila', and fancy chocolates with imposing French titles such as 'Le Roi de Chocolat'.

Hershey took frequent research trips to check out the competition. In New York in 1896 he met Catherine Sweeney, the young daughter of Irish immigrants, who worked in a candy store in Jamestown. The beautiful 'Kitty' proved to be an irresistible and charming confection, and Hershey's trips to New York took on a new urgency. The differences in their ages and backgrounds, and his mother's disapproval, counted for nothing. In 1898 Kitty Sweeney became Mrs Milton Hershey, and moved into his house in Lancaster.

Kitty was no Quaker wife. Unlike George Cadbury's wife Elsie, with her air of sobriety and her plain clothes modestly buttoned to the neck, Kitty had a sensational wardrobe crammed with the latest fashions. While Elsie was esteemed in Birmingham for her generous charitable works, Kitty was learning how to spend Milton's money for him, and her flirtatious style attracted comment. At least one Lancaster

Milton and Catherine ('Kitty') Hershey.

resident confided to being so taken by her beauty that she could not resist spying outside her house, hoping for a glimpse of the woman who embodied such an exotic contrast to the closed, religious community in which she now found herself.

Hershey indulged Kitty's desire to travel with long trips to Europe. There was a purpose to these visits: while his wife went shopping, Hershey went to see the British and Continental chocolate manufacturers for himself. He had heard about the chocolate works at

Bournville both through his burgeoning export trade and from admiring reports in fashionable American magazines.

Annie Diggs, a reporter for *Cosmopolitan New York*, took a tour of Bournville in 1903, and admired the effects of William Harvey's designs. 'The very streets of Bournville laugh in the face of crude conventionalism,' she declared. 'The monotony of capitalistic housing . . . with rows of all-alike houses is prohibited.' In its place were tree-lined walkways set out in curves and angles 'that follow the natural undulation of the land in all its native beauty'. She applauded the living conditions of the residents: 'Even the lowest priced of the cottages affords . . . attractive interior features, alluring gardens and an environment soul-satisfying to refined tastes – and all this at less cost than one clammy blackened room in a fever haunted city court where human creatures herd from birth to death.' She even praised the gardening: 'Why it is the very joy of life among the villagers! The men not being overworked in the factory go straight to their gardens with keen delight.' She caught sight of 'gloriously happy youngsters . . . skipping after their fathers . . . with spade and barrow to work their allotments after factory hours'. She described the recreation grounds, with 'charming woodland haunts', the 'fine pavilion for entertainments . . . cricket fields, football grounds, fishing pools and swimming places . . . reading rooms with the best of books . . . literary societies, debating clubs and

The lily pond in the women's grounds at Bournville.

institutions for serious study', not to mention 'professionals to instruct the girls in their gymnasium'.

At the chocolate works, Annie Diggs was impressed by the efforts to improve the health of the employees. No wonder, she observed, staff morale was high. Standing at the 'pretty lodge gate' as the factory closed, she saw 'a cheery, bright procession of working girls' who 'look fresh and dainty enough to grace any home or gladden any mother's heart'.

She found George Cadbury a passionate advocate for his scheme. 'We must destroy the slums of England or England will be destroyed by the slums,' he told her. 'We must give English children a chance to grow. We must not house our workers in a vile environment and expect their lives to be clean and blameless. We must do justice in the land.' For Annie Diggs, Bournville was 'a good dream come true' that 'invites duplication throughout England'. Indeed, she concluded, why stop there? 'Why not the United States as well?'

For Milton Hershey, Bournville was a model for the perfect business empire. When travelling to Europe with Kitty he almost certainly had a chance to visit Bournville for himself to see what Quaker philanthropy could achieve. From his carriage he would have been able to take in the neatly cultivated gardens and village green bordered by friendly clusters of houses, the shady streets named after trees, adding to the feel of a country haven: Willow Road, Oak Tree Lane, Selly Oak Road, Holly Grove – a little world set apart from the festering city. Richard had donated funds to help George's enterprise, and in the late 1890s thirty-three almshouses were under construction on the far side of the women's grounds. Placed in a quadrangle around a closely cropped lawn and backing onto a beautifully cultivated orchard, their atmosphere was reminiscent of centuries of English country village life. To Hershey, this idyll was the perfect community in miniature, a template for an entire society. He was inspired.

And there was something else to inspire him in Europe: milk chocolate. Through Lehmann's, he was introduced to several Swiss manufacturers. His own experiments at the Lancaster Caramel Company had shown him that it was not easy to make a chocolate bar, and he wanted to know how the Swiss created their superior milk chocolate. There are unconfirmed reports that while in Europe, Hershey attempted a little industrial espionage, although he himself always remained silent on this issue. Some claim that he was hired to work in Swiss chocolate factories without revealing who he really was. Other sources say that he lured staff away from Swiss firms by promising them better pay.

Whatever the truth, by 1899 the King of Caramel had made a radical decision: he would sell his caramel company. 'Caramels are just a fad,' he said, convinced his sales had peaked. 'But chocolate is a food as well as a confection. I'll stake everything on chocolate!' After prolonged negotiations, in August 1900 he received a cheque for $1 million for his Lancaster Caramel Company from Daniel Lafean, who was developing a rival caramel firm. Basking in his newfound status, the former push-cart vendor to whom no one would lend money was now a millionaire, occasionally seen cruising around in the first 'horseless carriage' on the streets of Lancaster.

Milton Hershey had a new dream. He would build a factory in a cornfield, and create the Hershey Chocolate Company, the largest chocolate works in the world. Like the Cadbury brothers, he would create a model town around his chocolate works. And he would build this American 'Bournville' where he himself had started out as a child, in rural Dauphin County, near Derry Church.

There was, however, just one obstacle. Hershey had the expertise to build a business, and the resources to create his chocolate town. But a key ingredient was missing. He did not have the recipe for Swiss milk chocolate.

PART THREE

11

Great Wealth is Not to be Desired

Richard Cadbury was a passionate supporter of the Egypt Exploration Fund, formed in the late nineteenth century to conduct archaeological excavations on sites in the Nile Delta that 'had rarely been visited by travellers'. In January 1897 he embarked on a tour with his family to see these hidden treasures, an experience of 'unclouded happiness', recalled his daughter Helen. Each day they were overawed by such wonders as 'the majestic figure of the Sphinx gazing across the desert . . . the solemn grandeur of Karnak by moonlight', and the constantly changing panorama from the Nile.

After touring Egypt, the Quaker party made their way to Palestine, Bibles in hand as a guide. They wanted to retrace the steps of biblical figures such as Joseph and Moses, seeing what they saw, feeling what they felt. When their carriage drove through the Jaffa Gate into Jerusalem, 'we were all quiet', wrote Richard's wife Emma. 'It seemed so strange and wonderful to be in the city of which from babyhood we had heard and read and sung.' Eagerly cross-checking descriptions in the Bible with the places they visited, they toured Bethlehem, Jericho, Nazareth and Damascus, and saw the deep blue sea of Galilee. The trip was so inspiring that Richard was soon planning a return visit.

On 2 February 1899, he and his family set sail once again for Cairo. 'I wish you could all see father,' Emma wrote home. 'He is most enthusiastic, taking rubbings and drawings wherever he can.' The family could not resist returning to the Pyramids, then visiting the Sphinx on camel-back. Someone took a photograph: a moment frozen in time. It shows a happy group laughing in the sunshine, the women in prim, high-necked blouses, long skirts and boaters; the men dressed like country squires – except for Richard. Happy and sunburned, he had turned native, and looked in his long, flowing robes as if he would never return to the land of grey skies.

Shortly after a trip up the Nile, the sixty-three-year-old Richard felt unwell. The doctor dismissed it as nothing of consequence; an ordinary case of 'Nile throat'. Richard did not fuss. He was eager to see Jerusalem again, and pressed on with the tour. By the time he arrived, after a two-day carriage ride from Jaffa during which he was unable to eat or drink, he was clearly more seriously ill.

Richard's decline was terrifyingly rapid. He grew weak and frail, and within three days he was slipping in and out of consciousness. Emma saw his eyes open wide, and 'brighten with a joyous surprise' as he 'gazed upon some glorious sight that was hidden from her'. In that moment she felt a frightening certainty that 'he could never come back to earth again'. During the night of 21 March, before she could summon the children to his bedside, Emma saw her husband die as quickly and quietly as a candle being blown out.

At Bournville, George had no idea that his brother was ill. It was a shock when the telegram with the news of his death arrived from Egypt. He had only just received Richard's happy letter from Abu Simbel: 'The sun had set when we anchored . . . so that we could only see in the shadow of the rocks a faint outline of the mammoth statues that guarded the celebrated temple of Rameses the Great . . .'

It was a terrible loss. As brothers and partners their lives had been bound to each other. The early days at Bridge Street were still vivid in George's mind, together with images of everything they had shared. The voice of his brother, so readily called to mind, was always positive. His ghost was everywhere. Together they had conjured up the dream of a utopia, and by sheer determination they had made it work. It was hard to believe that the brother, in robust health, who just a few weeks earlier had walked with him around the newly completed almshouses that he had donated to Bournville would never again be seen around the village. It took sixteen days to bring Richard's body home. Thousands congregated at Lodge Hill Cemetery in Selly Oak on the morning of 8 April 1899 to hear the simple funeral service. 'All of Birmingham has been thrown into mourning by the sad and startling news,' reported the *Birmingham Post*.

Over the spring and summer of 1899, the family catastrophe prompted a complete reorganisation of the chocolate factory. The Cadbury brothers had previously agreed that if either of them died, the firm would be reorganised as a private limited company, handing opportunities down to the next generation. Accordingly, Richard and George's four oldest sons found themselves joint managing directors of the vast enterprise. For the Cadbury cousins, it was a complicated inheritance.

Richard's oldest son, thirty-six-year-old Barrow, had a decidedly non-materialistic streak that sat incongruously with the directorship of a large firm. He was intensely committed to the Society of Friends, and to promoting its ideals such as peace and the unity of Churches. With his eye for detail, he was scrupulous in the management of Bournville's cashiers' department. He also delighted in introducing new technology: adding machines that printed on rolls of paper, typewriters and telephones. Soon he adopted the novel idea of installing internal telephones across Bournville. Barrow's brother, thirty-two-year-old William, favoured the outdoor life like his father. He already had ten years' experience at Bournville, running the engineering department and introducing machinery to keep pace with new product lines. Now he took responsibility for sales, and was involved in sourcing cocoa for the firm.

Their two cousins, George's oldest sons, had less experience in the family factory. Twenty-six-year-old Edward, a shy man known for his business acumen, was appointed to lead the fast-growing export department, and quickly made plans to check on the company's sales teams across the world. His twenty-one-year-old brother, George Junior, who had a passion for science, found himself with arguably the most nerve-racking assignment of all: creating new brands with which to challenge the Dutch and the Swiss.

For George Junior, it had been a 'shattering blow' when his father had sternly ordered him to leave his science studies at London University two years earlier to join the family business. Impatient and mischievous – as a child he had not been above such pranks as rowing the maids to the island in the lake at the manor and leaving them there – the burdensome role of joint managing director sat heavily on his young shoulders. To his father, the culture of further education was unimportant compared to 'the culture of the soul and earning a livelihood'. He felt that his son should apply his scientific skills to the needs of the firm's new chemistry department.

The Cadbury team had struggled for ten years to find a formula that could beat the Swiss milk chocolate, but without success. Their first attempt to be tested in the market, made with milk powder, was not launched until 1897, and was struggling to secure a foothold in the shops. Grocers preferred to stock Swiss chocolate, because their customers asked for it. Whatever combination of ingredients George Junior and his team tried in the laboratory, their milk chocolate remained stubbornly coarse, dry and unsaleable. Rumours were rife that both the Rowntrees and the Frys were preparing to launch a milk chocolate brand.

George Junior knew he had a young rival at Rowntree's factory in Haxby Road in York. Like George Cadbury, Joseph Rowntree had turned his chocolate works into a limited liability company and passed the baton to the next generation. His second son, twenty-eight-year-old Seebohm Rowntree, was running the research laboratory, and he too had been tasked with developing products to rival the Swiss. For George Junior there was no knowing what the Rowntrees might come up with. Joseph Rowntree's nephew Arnold had already proved himself in an area his uncle had deliberately avoided, with what Joseph thought of as advertising 'stunts'. First there was the motor car – at a time when such a thing was still a novelty – with a gigantic tin of Rowntree's Elect Cocoa attached, which rattled disconcertingly and demanded attention as it was paraded through towns. Then there was the Oxford and Cambridge boat race of 1897, when Arnold had the temerity to cover a barge with posters for Elect Cocoa and sail it right across the course. It was not something his uncle would have done, but it was yielding results: sales of Elect Cocoa were taking off. As for Fry, it was anyone's guess what it might conjure up next.

After a short apprenticeship at chocolate factories on the Continent, George Junior secured an invitation to tour Peter's plant in Vevey. This prompted him to set up a specialist milk-condensing plant at Bournville to investigate the best ways to evaporate milk in bulk without spoiling it. On outings every weekend in his shiny new Lanchester car – the latest model, with a roof – George Junior could see only too plainly the dispiriting lack of success of Cadbury's first milk chocolate bar to reach the shops. It was a disaster. The Swiss sold thirty tons of milk chocolate a week in Britain, while Cadbury could not manage even one ton. But what George Junior came up with next would transform the fortunes of the company.

In Bournville, George Cadbury Senior, while still chairman of the company, had stepped back from the day-to-day operation of the business. He was keen to use his time to expand his philanthropic interests. This was something he had discussed with his wife, and they had many plans.

Elsie had proved to be the perfect wife for George, capable, warm-hearted and totally committed to Quaker principles. The manor resounded with the exuberant sounds of their growing family: by 1899 there were ten children from George's first and second marriages. Portraits of the time capture the strength of the Victorian family. George, bearded and smiling beneath his top hat, with Elsie seated at his side. The teenagers are grouped behind them, the babies sit on their laps, and the younger children are arranged at their feet. The girls have long, flowing hair, long skirts and frilly white blouses; the boys are dressed like their father, in smart dark suits and ties. It is a portrait of success, but not merely material success. Both parents believed that children must have a firm foundation from which to explore the world.

Despite the formality of the family pictures, Elsie was not one for stuffy restrictions. The children were left with their governesses during the day, while Elsie invariably went on a round of public engagements. When she returned one day to find that her daughters Dolly and Molly had been sent to their rooms as a punishment for the most unladylike act of climbing on the roof, she let them out at once. All children should be allowed to take risks, she said, otherwise they would have no chance to learn. Elsie was unfailingly zealous in her commitment to good works, even if it meant putting her holiday in jeopardy. When it was time for the family's holidays, it was not uncommon for George Senior, who liked to arrive thirty minutes early for a train, to go on ahead to the station with the children in the pony and trap. Typically Elsie, working for the poor until the last minute, would dash up as the train's whistle blew. 'My recreation begins the moment I drop into a comfortable railway carriage,' she told a friend, 'having counted my family to see that none is missing.' On Sundays she played the organ installed in the oak room on the manor house's ground floor, where the family gathered to sing hymns. George doted on her, and if they were apart they wrote to each other sometimes twice a day. 'What a thing it is to be ruled by one's wife,' George said.

Each year they threw open the grounds of the manor house for parties of children from some of the roughest districts of Birmingham. They built a large hall known as 'The Barn' in the park to provide tea and refreshments for up to seven hundred children. George Senior, with his love of nature, believed strongly that every child should have access to playing fields in clean air. Games were organised in the open fields, but invariably the star attraction was the open-air baths. More than fifty children could bathe at one time, and for the young visitors, most of whom had no access to a bath, the experience was a revelation. With the sun on their backs, the sparkling water always inviting, the boys from the inner city had no desire to leave and would stay in the water all day, until they were blue and shivering and cleaner than they had been in years.

The parties at The Barn were just an informal beginning. George and Elsie had both witnessed the critical problems of urban industrial living: housing shortages, crowding, poverty and the social problems that come with deprivation. George wanted to take a scientific approach to tackling these problems, and aimed to use Bournville as a testing ground for reform. But before he could implement his plans, he was confronted by some difficult questions about how a man should best use his wealth. On 11 October 1899, an ill-equipped army of 35,000 Boers prepared to take on the British Empire, the second time tensions in southern Africa had erupted in conflict in less than ten years. The Boers, a group of farmers of Dutch descent, had migrated inland to the Orange Free State and the Transvaal, territories that were rich in mineral wealth, to escape British rule. But there were many who agreed with the British mining magnate Cecil Rhodes that the British Empire should command the continent of Africa 'from the Cape to Cairo'.

George and Elsie were shocked at the way the press fuelled the appetite for war. The brash new *Daily Mail*, under such rousing head-lines as 'For Empire and Liberty', was bombastic about British pros-pects. 'Brain for brain, body for body,' the paper assured its 750,000 readers, 'the English speaking people are much more than a match for the Dutchman.'

George Senior was approached by a rising star of the Liberal Party, the radical Welshman David Lloyd George. He was opposed to the war, and knew that George Cadbury shared his views. Lloyd George had a challenging proposal. The national press was speaking almost with one voice in favour of war, and fuelling the jingoistic fervour of its readers. Lloyd George was keen to ensure that the public heard alternative views. Very few papers dared to take on the establishment, challenge those who advocated war or probe the interests of the British mine owners in southern Africa: the *Manchester Guardian* in the north, and London's *Morning Leader* in the south, were the only papers that consistently opposed the war. Lloyd George had a special interest in the *Daily News*. This once-radical newspaper had been founded by Charles Dickens in 1846, and had championed liberal reforms and social issues; now it was taking an editorial line that supported the war. He asked George to join a syndicate to buy the paper.

George was sympathetic to Lloyd George's plan. He believed that British diamond speculators and mine owners in southern Africa like Cecil Rhodes wanted to suppress the Boer government in the Transvaal, and keep control of the mines for themselves. He abhorred the greed and imperialism masquerading as a just cause, and like Lloyd George believed that the cost of the war was delaying social reforms at home. But he hesitated. The ownership of a national newspaper would be a completely new venture. He had previously avoided taking a prominent role in politics, turning down an invitation from William Gladstone in 1892 to stand as an MP. George wrestled with the decision. What was the best way to use his wealth to benefit society at large? Should he seek to influence and educate public opinion and present issues honestly through a national newspaper? Or should he instead concentrate on developing his template at Bournville?

Finally he told Lloyd George that he was reluctant to take on the *Daily News*, but he could make a small contribution to his campaign by paying for an early train to take copies of the *Morning Leader* each day to key towns between London and Sheffield, so the public would be exposed to an alternative view.

George's greatest priority remained Bournville. Ever mindful of the corrosion to the soul that great wealth might bring, he took the step, with his wife's complete support, partially to disinherit his own children. There is a photograph that marks the day, on 14 December 1900, that George Cadbury Senior gave away his wealth. His original Quaker parish of sixteen homes clustered around the chocolate works had blossomed into an idyllic English village with 370 cottages and five hundred acres of land. Now he wanted to give it away to create the Bournville Village Trust. A large and solemn group had gathered in front of the Friends Meeting House on the village green to hear what he had to say.

'I am not rich as an American millionaire would count riches,' George Senior declared. 'My gift is in the bulk of my property outside of the business . . . I have seriously considered how far a man is justified in giving away the heritage of his children and have come to the conclusion that my children will be all the better for being deprived of this money. Great wealth is not to be desired and in my experience it is more of a curse than a blessing to the families that possess it.'

He explained that six of his ten children were of an age to understand how this action affected them 'and they all entirely approve'. Provision had been made for 'an insurance of a modest competence' for each child, but beyond this, it was up to them. Judging by the sober faces of the large family gathered around him, the gravity of this decision and the consciousness of his high expectations of them were all too plain.

George Cadbury was the first English chocolate entrepreneur to create a trust of this kind, and his hopes for what it could accomplish are clear from the Deeds. The aim of the Bournville Trust was 'the amelioration of the conditions of the working class and labouring population', with a special emphasis on improving their quality of life with 'improved dwellings with gardens and open spaces to be enjoyed therewith'. The homes were intended to be for a cross-section of society, and this was reflected in their prices or rents. At the time of the gift, 143 houses had already been sold to tenants on 999-year leases – some for as little as £150. The remaining 227 were let for varying sums, depending on their size. Every house had a garden that was sufficiently large to produce around two shillings' worth of fruit and vegetables per week – worth roughly £375 a year today – increasing its value to the tenant still further.

George wanted the village to maintain its quality as it grew. Once necessary repairs and maintenance had been carried out, the trustees were to use any remaining funds to buy land and build more homes, applying the same generous ratios of parkland and gardens to buildings. He wanted to prove that the scheme was economically viable, so that other philanthropists or investors could see its benefits and follow suit. If investors could make a financial return on providing quality housing for tenants of varying backgrounds, they might be inspired to copy the Bournville plan and thus contribute to the betterment of society at large.

The experiment at Bournville was part of a wider debate concerning the problems of inner cities. In 1898 a parliamentary shorthand writer, Ebenezer Howard, had published *Tomorrow: A Peaceful Path to Real Reform*, which set out a grand scheme for a revolution in town planning. Howard was concerned at the continuing exodus from the country to the towns: London's population had almost doubled in size between 1870 and 1900, from 3.9 million to 6.6 million, bringing with it all the familiar problems of urban poverty. Howard believed the solution lay in a futuristic scheme in which, he said, 'town and country *must be married*' (his italics). He toured Bournville, and believed the garden city idea could be developed to provide an entire social and economic system with the potential to tackle some of the key problems of late-Victorian capitalism.

Howard's idealistic scheme envisaged large areas of land converted to clusters of garden cities. Each one, of approximately 32,000 people, would be geometrically arranged, with green belts separating it from the next. Unlike Bournville, Howard's garden city would be arranged on a circular design, with boulevards and avenues surrounding a central park. This compact design gave it a human scale, and put most essentials within walking distance. He tried to anticipate every need, even foreshadowing today's shopping malls with an elegant glass arcade in each town called a 'Crystal Palace'. Howard hoped that workers would unite to help realise his vision, 'for the vastness of the task which seems to frighten some . . . represents in fact the very measure of its value to the community'. He was so passionate about his beliefs that he launched the Garden City Association in 1899 to put them into practice.

George Cadbury supported Howard's vision, which coincided with

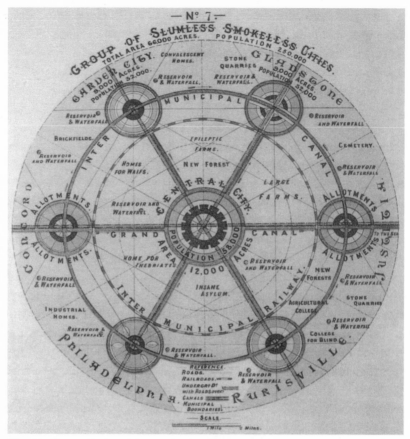

Ebenezer Howard believed his Garden City concept could help to solve the problems of the inner cities.

his own ideas on land use. He believed that injustices in the owner-ship of land 'lay at the root of many social evils'. If the estimated nine million households in England were accommodated in cottages at ten to the acre, he reasoned, they would occupy a mere 900,000 of the seventy-seven million acres in Britain. Since tests at Bournville showed that one acre of cottage gardens yielded twelve times more produce than one acre of pasture, this would be a much more effec-tive way of meeting the nation's food needs. A massive increase in death duties was needed, he thought, so that it would no longer be possible for most of Britain's land to accumulate in the hands of a few wealthy families. Bournville proved that land was more effectively cultivated 'in the hands of the people themselves'.

As for the difficult decision to disinherit his own children, George set out the reasoning behind this six years later, in an interview he gave to a committee of the Canterbury Convocation investigating social problems, including the accumulation of wealth. At the heart of it, he explained, was his unwavering faith. He believed that 'Every man must give an account of himself to God.' According to Matthew 19:24, 'It is hard for a rich man to enter the kingdom of heaven . . . it is easier for a camel to go through the eye of a needle.' For George Cadbury Senior it was simply wrong that a lucky few 'have a supera- bundance of wealth . . . and every conceivable comfort and luxury', while countless others in 'so-called Christian lands . . . lack things which are essential to health and morality'. Great wealth was of questionable value. 'I have seen many families ruined by it, morally and spiritually,' he told the committee. And while money brought no lasting happiness to the rich, it 'certainly brought a vast amount of misery upon those who did not have their share'.

George and Elsie continued to contribute their personal wealth to develop the community at Bournville. Over the years several hand- some public buildings were built around the village green. The first was a meeting place known as Ruskin Hall, which eventually became the Bournville School of Arts and Crafts. It provided professional qualifications such as teacher training, as well as craft skills like dressmaking and metalwork. They also paid for the Bournville Schools, and spent happy afternoons together perfecting the designs that equipped each school with a modern library, scientific laboratory and extensive playgrounds. And of course everything had to be in a beau- tiful, timeless setting. When staying in Bruges in Belgium, George and Elsie had been enchanted by the sound of the cathedral's fine bells, and arranged to make an exact copy of them for Bournville. It was a great day when all twenty-two bells were hoisted into position below the domed cupola in the tower of the Infant School, and the carillon rang out around the village green.

The Cadbury brothers had initiated a great many schemes targeted at improving the health and well-being of their staff, and George was dedicated to developing them further. A job at the chocolate factory was not for the physically faint-hearted. The men's sports grounds now extended to twelve acres, and there were Bournville teams for every conceivable sport. In the summer months, the outdoor pool for men was popular. An imposing indoor pool was completed for women in 1905, and thousands of employees would learn to swim. The women also had twelve acres of grounds for sport. Fitness training was compulsory for anyone under eighteen – an hour a week was set aside during work time. Village events were organised on the green, including a maypole for children's dancing in spring, summer fêtes and sports days.

In 1902 a doctor was hired for the staff, and the medical department expanded over time to include four nurses and a dentist, all of whom, like the doctor, were available free of charge. Free vitamin supplements were provided for those who required them, and a convalescent home in the Herefordshire countryside was built for staff in need of a rest. While all this may seem quaintly paternalistic now, at a time when workers were often subjected to unhealthy or even dangerous working environments, it is small wonder that people were queuing up to be employed at Bournville. At the turn of the twentieth century George Senior took a further step and asked his son Edward and his nephew Barrow to look into the novel idea of creating a pension scheme.

There were the spiritual needs of his community to consider as

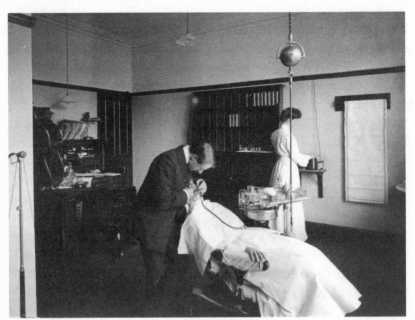

The Bournville dentist, 1905.

well. George believed it was not possible to grow morally or spiritually in a slum. Only in the open spaces of the country was it possible for a man to come into touch with nature, 'and thus know more of nature's God'. At Bournville, apart from the Friends Meeting House, a site was found for an Anglican church, a village hall and a vicarage. George welcomed worshippers of different faiths. He was a friend of William Booth, a Methodist, who had founded the Salvation Army in 1865, and he valued Booth's message of a 'practical religion' encouraging members to work in the slums. George believed that all Churches should unite to tackle issues such as helping the poor, and created a central library so ministers and preachers could share works from different faiths.

George and Elsie's efforts to help the underprivileged were showing tangible results. In 1919 studies were carried out that compared children aged six to twelve who had been brought up in the poor Floodgate Street area of Birmingham with children of the same age who grew up in Bournville. The Bournville children were on average two to three inches taller and eight pounds heavier than their Birmingham counterparts. Infant mortality in Birmingham was 101 per thousand births, while in Bournville this figure was halved.

Above all, George wanted the community at Bournville to improve the lives of children from Birmingham's slums. This may well have been due to the enduring influence of his mother, Candia, who in her temperance visits to the poor districts of the city had been concerned for the innocent young victims caught in the cycle of deprivation. George Senior revisited those same teeming, dirt-encrusted inner-city areas, where a dozen families might occupy one house. In his ideal world, no child would be brought up in a place where a rose could not grow. He arranged to buy sites from Birmingham City Council and turned them into playgrounds, hoping that other wealthy families in the city would follow his example. Ideally, he argued, there should be a playground every four hundred yards, so that children throughout the city had daily access to spaces where they could play and grow more healthily.

At Bournville, George and Elsie created The Beeches, a large house with extensive grounds which was used as an invalid home in the winter, and in the spring became a camp where children from the industrial slums could enjoy a two-week holiday. Under the jurisdiction of the endlessly capable Mr and Mrs Cole, thirty children stayed at a time. They could eat as much as they wanted, sleep in clean beds and roam the garden and the surrounding fields all day. According to contemporary reports, the children enjoyed their stays so much it was not unusual for them to be 'unaccountably missing' on departure day. They were invariably found hiding under beds or in cupboards. The children were weighed on arrival and departure, and were found to be two to three pounds heavier after their visits.

Possibly inspired by his father's campaign to help the chimney sweeps, some of whom were badly maimed by their work, George Cadbury also had a scheme for children who were unable to play. He knew that many of the cripples on the streets were the luckless victims of 'cruel circumstances or ignorant chance'; especially tragic were those who had been harmed by 'the carelessness of their own drunken parents'. George bought a grand house called Woodlands, which was set in six acres of parkland just across the road from the manor house, and had it adapted to cater for the needs of crippled children.

George enjoyed making the rounds of these charities. He liked to visit The Barn when a party was in progress, and according to his biographer Alfred Gardiner he visited the handicapped children at Woodlands each week, a 'large box of chocolates tucked under his arm'. Invariably his entrance was greeted by with 'loud shouts'. After chatting downstairs with those who were recovering their mobility, he would go upstairs to see the long-stay children, who had more serious injuries. He spoke to each child, handing out gifts, and employed a surgeon to do anything that could be done to improve their chances in life.

George Cadbury's religious convictions shaped his world, and unified every aspect of his life. His faith was simple and unquestioning, and emphasised practical help to those in need. Walking around Bournville, he could see the results of his endeavours in the houses and public buildings where there had once been only muddy fields. His and Richard's improbable dream had been turned into solid bricks and mortar, into something powerful for good. And it was all from chocolate. Chocolate, or rather the humble cocoa bean, had created a little Eden in the English Midlands. He hoped the success of his charities and trusts would inspire others. 'Example is better than precept,' he said. 'If you can show that your life is happier by giving [than by hoarding], you will do more good than by preaching about it.' To his great pleasure, in 1900 he was approached by Joseph Rowntree, who told him that he too wanted to create a model village.

◆◦ʃ❀ʒ◦◆

In York, Joseph Rowntree, like George Cadbury, found it liberating to delegate much of his business to the younger generation. He too had an all-consuming interest in tackling the problems of poverty, a passion shared by his son Seebohm. Both Joseph and Seebohm had been much moved by a powerful series of books on *Life and Labour of the People in London* written by Charles Booth during the 1890s. Booth popularised the idea of a 'poverty line' – a minimum weekly sum that was needed to maintain a man and his family at a basic level. His study highlighted the levels of squalor and degradation that London's poor endured, and he argued that the state had the means to help them. Seebohm Rowntree wanted to apply a dispassionate scientific analysis to the causes of poverty, and despite his duties at the chocolate works, he found time to embark on a groundbreaking study.

Seebohm chose York as a representative provincial town, and set out to gather data on 11,560 families living in 388 streets – two-thirds of the town's inhabitants, including, he said, 'the whole of the working class population'. He and his investigators undertook sensitive enquiries door to door, questioning people on their rent, income, the number of residents in each household, access to a water tap, diet and other personal information.

The page after page of case notes that resulted form a vivid snapshot of poverty in York at that time. Each distils the bare facts of a heart-rending family struggle. But how could Seebohm use this information to analyse poverty more systematically?

Labourer. Married. Two rooms. Four children. Chronic illness. Not worked for two years. Wife chars. Parish relief. This house shares one closet and one water tap with eight other houses. Rent 1s 7d . . .

Woodchopper. Married. One room. Parish relief. Wife blind. Mostly live on what they can beg. This house shares one closet and one water tap with two other houses. Rent 2s . . .

*Labourer. Married. Four rooms. Six children. Filthy to extreme. This
house shares one closet with another, and one water tap with five
others. Rent 3s 6d . . .*

*Husband in asylum. Four rooms. Five children. Parish relief. Very
sad case. Five children under thirteen. Clean and respectable but
much poverty . . . This house shares one closet with another house
and one water tap with three other houses. Rent 3s 9d . . .*

*Chimneysweep. Married. Two rooms. Five children under thirteen.
All sleep in one room . . . Man in temporary employment earning
2s a day . . . a bad workman . . . incapable of supporting his family
decently . . .*

*Polisher. Married. Four rooms. Two children. Parish relief. Wife washes
. . . Man not deserving; has spent all large earnings on drink. Fellow
workmen have made several collections for him. All speak badly of
him. House very dirty. Rent 3s 10d . . .*

After consulting with nutritionists and other experts on the
minimum requirements for a basic diet, Seebohm Rowntree set a
poverty line for a family of five at 21s.8d – roughly £75 today – a
week, acknowledging that this allowed for a diet 'less generous . . .
than that supplied to able bodied paupers in workhouses'. He defined
'primary poverty' as those living below this poverty line. No matter
how carefully they spent their money, this group did not have enough
to cover the minimum basic needs of life, which Seebohm described
as 'the maintenance of merely physical efficiency'. Incredibly, 7,230
people in York, almost 10 per cent of the city's population, fell into
'primary poverty', lacking the means to feed themselves adequately.
    Seebohm's analysis showed that low wages were the biggest single
cause of 'primary poverty'. Half the men in this category had jobs,
but they received such a pittance that they could not meet their
families' basic needs. The other key causes of primary poverty were
the size of the family and the death or illness of the wage-earner.
    Seebohm next analysed the 13,072 people suffering what he called
'secondary poverty' – roughly 18 per cent of York's population. To

the investigators, this group appeared just as poverty-stricken as the first, despite the fact that its members should have had enough money to meet their basic needs. The reason they were failing to do so was that they were spending some of their income on non-essentials such as drink. Seebohm found that a number of other factors, such as inadequate housing or overcrowding, were also contributing to their plight. Taking the two groups together, he showed that 27 per cent of the population of York were living in either primary or secondary poverty.

Seebohm Rowntree was 'much surprised' that his findings coincided closely with those of Charles Booth. Booth had estimated that 30 per cent of Londoners lived in poverty. If the findings in London and York could be extrapolated to the rest of Britain, wrote Seebohm, 'we are faced with the startling probability that from 25–30% of the town populations of the UK are living in poverty', something that he thought should prompt 'great searchings of the heart'. Surely, he asked, 'no civilisation can be sound or stable which has at its base this mass of stunted human life?' For him it was unacceptable that 'multitudes of men and women are doomed by inevitable law to a struggle for survival so severe as to cripple or destroy the higher parts of their nature?' His research was published in 1901 as *Poverty: A Study of Town Life*. One of those who read it was Winston Churchill, a Tory MP at the time. Rowntree's study, he declared, 'fairly made my hair stand on end'.

Joseph Rowntree was keen to do what he could to alleviate the situation his son had revealed in York. By 1900 he was at last in a position to do this: in the five years since moving to Haxby Road his business had doubled in size, with annual sales approaching £500,000. Staff numbers had risen to 1,600, and he knew that at least some of them were included in his son's study.

After discussions with George Cadbury, Joseph purchased 150 acres of land three miles outside York, and hired an architect to design a village on parallel lines to Bournville. He envisaged a similarly idyllic community, a utopia of cottage gardens resplendent with produce, where children from the city could play freely, taking on the colour of health. He wanted his village to be affordable for the poorest slum-dwellers of York, many of whose families were surviving on less than £1 a week. When the lowest rents of around five shillings proved to be beyond their means, Rowntree commissioned simpler cottages to be built, without bathrooms or hot water, for £135 each, and let for four shillings per week.

It took time for his experiment to come to fruition, but gradually, with the addition of a village hall, schools and playing fields, the pretty garden village of New Earswick took shape. In 1904, at the age of sixty-eight, Joseph handed over the estate to the non-profit Joseph Rowntree Village Trust. To complement the work of the Village Trust, he also created the Joseph Rowntree Charitable Trust and the Social Services Trust. These had the remit to investigate social and religious issues, and other matters of 'importance to the well-being of the community'. He committed half of his total wealth to the three trusts.

The discussions between Joseph Rowntree and George Cadbury brought the two friendly rivals together on another issue of keen shared interest. Joseph's older son, John Wilhelm, was beginning to question the Quaker movement. The Society was suffering from a decline in numbers, which he felt was caused by its restrictive codes of practice. He believed that it was stagnating, and was in danger of becoming 'little more than a hereditary social club'. Where was the spark? What was its mission? John Wilhelm organised a series of meetings to discuss ideas, and called for the creation of a permanent college that would develop Quaker thinking. George Cadbury offered his former home at Woodbrooke, on the outskirts of Bournville, for this college, the only one of its kind in Europe. He and Elsie wanted Woodbrooke College to become a retreat where the spirit might be recharged, and funded scholarships for students from around the world. They were hoping for a rebirth of the faith, after centuries of obedience to now-outdated ideas.

Joseph Rowntree was not the only entrepreneur in England who was inspired to copy Bournville. Early in the twentieth century, Sir James Reckitt, a Quaker and a successful businessman in household goods, developed a similar garden village in east Hull. Ebenezer Howard's ideals began to take shape as he founded his first garden city, at Letchworth in Hertfordshire, in 1903, followed by Welwyn Garden City. In 1907 a friend of George Cadbury, Henrietta Barnett, created Hampstead Garden Suburb in north London.

Soon there was a stream of visitors to Bournville from overseas. A

Frenchman, Georges Benoît-Levy, returned home to create a garden village at Dourges in northern France. In Germany, Margarethe Krupp, who inherited a fortune from her husband's large armaments firm, gave a million marks to build a big estate at Margarethenhohe in Essen, on condition that the architects studied Bournville before drafting their plans. Word of Bournville's success reached across the Atlantic, and to India, China and Australia. 'Bournville,' declared the *Melbourne Age*, 'is as important to England as a Dreadnought.'

◄❀►

While George Cadbury Senior was happily absorbed in creating his model community at Bournville, he could not ignore the wider issues in society. The deepening crisis of the Boer War provoked him into action. In the early autumn of 1900, a rector's daughter from Cornwall, Emily Hobhouse, a member of the South African Conciliation Committee, heard rumours of a horrific new practice being used by British commanders. Boer women and children were being held in a 'concentration camp' – so named for the policy of 'concentrating' them in one location – allegedly for their own protection. She set sail for southern Africa and soon revealed that there was not one, but thirty-four concentration camps. Worse, it was a mockery to call these 'refugee camps': the conditions in them were so barbaric that they were more like death camps. She denounced the system as 'hollow and rotten to the core'. But when she returned to Britain she was dismissed by the jingoistic press as 'a rebel and a liar and an enemy of the people'.

David Lloyd George had not given up on his efforts to find a Liberal backer for the *Daily News*. George Cadbury, shocked by developments in the war, the corruption of the mine owners and the recent revelations of concentration camps, began to see that owning a national newspaper could have a value. It seemed a matter of duty to use his wealth to influence public opinion. 'This war seems the most diabolical that was ever waged,' he told the Independent Labour MP John Burns. 'Just now it seems to me that speculators, trust mongers, and owners of enormous wealth are the great curse of this world and the cause of most of its poverty!'

In 1901 he agreed to Lloyd George's proposal, and put up £20,000 to join a syndicate to purchase the *Daily News*. By any standards, this was a colossal sum: enough to build more than eighty new houses at Bournville. But in sharing ownership of a national paper he hoped to expose other social ills, such as inhumane factory conditions. He would have a national voice that could promote the Quaker ideal of pacifism, and speak for the inarticulate and the unfortunate.

Opposing the war was no light matter. Advertisers responded swiftly by removing their business from the *Daily News*, and the paper's losses soared. By the end of 1901, the other leading investor, a Mr Thomasson of Bolton, wanted out. George faced a tough choice. He could sell his share, and run the risk that the paper would be bought by someone in favour of war. Or he could put up another £20,000 and buy the paper outright. Despite its rising losses, he chose the latter course of action.

As sole proprietor, George was able to appoint an editor who shared his views: Alfred George Gardiner, who would later write his biography. Under Gardiner's editorship, the *Daily News* opposed the war, and also drew attention to the scandal of tens of thousands of Chinese coolies labouring in inhuman conditions in South African mines. With provocative headlines such as 'Yellow Slavery', the paper condemned the Tory government for condoning the practice in order to support wealthy British interests.

Gardiner and his team also highlighted the urgent need for labour reforms at home. The *Daily News* was tireless in its exposure of appalling working conditions in Britain. The paper funded an exhibition in London that revealed the shocking exploitation of those working in sweated labour. It found women making shirts in their homes for less than a penny an hour, repairing sacks for two shillings a week, or chain-making for six shillings a week, often working more than twelve hours a day. George Cadbury became President of the newly formed Anti-Sweating League, and was supported by the indefatigable efforts of his oldest son, Edward. Edward wrote two books summarising the League's findings: *Sweating*, which highlighted the need for a minimum working wage, was published in 1907; and *Women's Work and Wages* followed in 1908. The *Daily News* also campaigned for unemployment benefits and old age pensions, and Edward and his father helped to create the National Old Age Pensions League to champion the cause of state support for the elderly. It appeared that the dream of using the wealth created by the family chocolate business to promote social reform and justice was finally coming to fruition.

George Cadbury was finding his thinking increasingly in step with the emerging Labour movement, which championed the interests and needs of the working population. He agreed with the trade unions that campaigned for improved labour conditions, shorter working hours and better provisions for workers, such as sickness benefits: these were steps he had already taken at Bournville. But in the 1900 general election the Labour Representation Committee – a precursor of the Labour Party – won only two parliamentary seats. 'We want a hundred working men in Parliament,' George declared. 'Only then will the condition of the people become a living issue.'

He was finding himself in growing conflict with the *Daily News*'s chief executive, the journalist Thomas Ritzema, who had strict Puritanical views. Not content to promote pacifism and labour causes, the moralistic Ritzema cancelled the racing pages and the betting tips, and refused to accept any advertisements for alcohol. Circulation plummeted further as the paper began to be seen as needlessly moral and censorious. Soon the *Daily News* was costing George up to £30,000 a year. His fortune, cultivated with such prudence and exactitude, was being drained. Worse was to come.

George began to hear appalling reports of slavery in Africa. These suggested that the practice was rife in the very plantations where he was buying most of his cocoa: on São Tomé. This was a moral blow more severe than any business setback could have been.

12

A Serpentine and
Malevolent Cocoa Magnate

The first warnings came in 1901, when George's nephew, the thirty-four-year-old William Cadbury, sailed across the Atlantic to visit one of the company's small cocoa plantations in the West Indies. As a leading buyer for Cadbury he had bought two small estates in Trinidad four years earlier to research improvements in methods of cultivation. William thrived on the outdoor life, and always eagerly anticipated his annual trip to the Caribbean. But this year he learned of troubling news.

Drying cocoa in Trinidad, c.1896.

The growers in Trinidad told him of a rumour they had heard about cocoa plantations thousands of miles away on the other side of the Atlantic. It concerned the two islands that had been the first places to cultivate cocoa in Africa: São Tomé and Príncipe, in the Gulf of Guinea. The Trinidad growers believed that some of the workers on these West African islands were slaves. William was worried. In 1900 Cadbury had bought 45 per cent of its beans from São Tomé and Príncipe. He knew very little about the islands, except that the beans cultivated there were superior to any others in Africa, and their production was prolific.

By chance, later that spring the Cadburys were notified of a plantation for sale on São Tomé. As William read the sales brochure, to his alarm he came across further evidence: included in the list of assets were 'two hundred black labourers worth £3,555'. 'The suggestion behind this statement was obvious and disturbing,' he wrote. He took the matter to the Cadbury board. 'This seems to confirm other indirect reports that slavery . . . exists on these Cocoa estates,' the minutes for 30 April 1901 record. The board asked William Cadbury to investigate.

The two islands were under Portuguese control, but slavery had been officially abolished in Portuguese colonies during the 1870s. The abrupt end to the slave trade on São Tomé had come in 1875, when 6,000 desperate labourers simply walked out of the plantations and entered the capital, demanding that they be treated like freedmen. How was it possible that any plantation owner could continue this grotesque practice? William Cadbury turned to Travers Buxton, the Secretary of the British Anti-Slavery Society, for advice.

He learned that the Anti-Slavery Society had received several accounts relating to the islands from missionaries and explorers in the years since 1875. A Swiss traveller, Heli Chatelain, reported seeing slaves in Angola in 1891 who were destined for São Tomé. 'Some of them looked healthy,' wrote Chatelain. 'The majority showed signs of bad fare; some . . . were starved to skeletons.' A French traveller in 1900 also observed slave gangs in Angola. 'All this trade is done with the protection of the Portuguese government,' he claimed in the *Anti-Slavery Reporter*. In 1902 William was introduced to a Scottish missionary, Matthew Stober, who had recently returned from central Angola. He too claimed to have witnessed the slave trade at first hand.

William was sufficiently troubled to set out in 1903 to Lisbon, to meet the Portuguese authorities and plantation owners for himself. Stober, a fluent Portuguese speaker, accompanied him, so that he could describe the horrors he had witnessed. In a letter to Joseph Storrs Fry II in Bristol, who was also buying São Tomé cocoa, William set out his concerns. He wanted to enlist the support of the other leading cocoa buyers, to bring the practice to an end.

As William and Stober made their way by train and carriage across France and Spain towards Lisbon, they had plenty of time to ponder the situation. Quakers had led the campaigns against slavery for three centuries. How was it possible for leading Quaker firms to be apparently involved in a barbaric trade that officially no longer existed?

From the time of the Quaker founder, George Fox, in the seventeenth century, Friends had believed in the sanctity of human life, and the significance of every individual in the eyes of God. In 1657 Fox expressed his unease about the slave trade in a letter to 'Friends beyond the Sea that have Black and Indian slaves'. He was so concerned about the matter that in 1671 he travelled to the Caribbean and America, where he spoke out against this cruel practice that reduced men to little more than cattle to be bought and sold. In 1727, the Yearly Meeting of Friends warned all members 'to avoid being in any way concerned in reaping the unrighteous profit arising from the iniquitous practice of dealing in Negro or other slaves'. Quaker businessmen on both sides of the Atlantic were advised by their elders to be wary of finding themselves unwittingly supporting slavery.

As slavery flourished, so did the Quaker denunciations. The trade stood in opposition to everything a Quaker stood for. This 'Hellish practice', declared Benjamin Lay in 1736, was a 'filthy sin . . . the greatest sin in the world'. His impassioned denunciation of anyone involved in the slave trade eventually obliged him to leave his role as a merchant in Barbados to settle in Abington, Pennsylvania, where he continued to preach. Slavery was 'of the very nature of Hell itself, and is the belly of Hell'.

Revulsion at the trade triggered a high-profile Quaker campaign for reform in the latter half of the eighteenth century. In England, Quakers were not permitted to stand as MPs themselves, but this did not stop them taking the first anti-slavery petition to the British Parliament in June 1783. The politician and philanthropist William Wilberforce was horrified at what he learned of the depraved trade, and took up their cause, becoming one of Britain's most prominent abolitionists. In 1787 when the Abolition Society was formed, the majority of its founders were Quakers. They devised ways of exposing the cruelties of the trade, such as publicising drawings of slave ships, showing how slaves were crammed into the transports shoulder to shoulder. Their campaign helped to pave the way for the British Slave Trade Act of 1807, which made it illegal to capture and transport slaves across the British Empire.

But this did not bring an end to the highly profitable trade. The Royal Navy intercepted more than 1,600 slave ships between 1808 and 1860, and liberated 150,000 slaves found on board. Any captain caught carrying slaves faced a fine of £100. It was not uncommon for slaves to be thrown overboard to avoid discovery. Quakers continued to campaign for abolition, creating the Anti-Slavery Society in 1823. They wanted not only to stop the trade, but also to free all existing slaves. Their work culminated in the Slavery Abolition Act of 1833, which led to the gradual emancipation of all slaves across the British Empire. After this, the British and Foreign Anti-Slavery Society took its campaign to other countries.

Many of William Cadbury's forebears, including his great-uncles Joel Cadbury and Benjamin Head Cadbury, the older brothers of his grandfather John, had been active anti-slavery campaigners. Joel Cadbury, who had emigrated to Philadelphia, witnessed appalling examples of slavery in America. On one occasion, during a business trip to New Orleans, he saw a large crowd gathered around a shed, and realised that slaves were being sold inside. As he entered, he saw a black woman on sale. Other slaves, herded into position like animals, awaited their turn. Almost naked, the woman was made to stand while the auctioneer pointed out her selling points. She was treated like a mere exhibit, described as healthy and fit for anything. When sold she would be the owner's property, to do with as he wished. Joel was so sickened he had to leave. His wife Caroline shared his concerns, and in 1860 became Treasurer of the Shelter for Coloured Orphans. The English

Quakers of Pennsylvania were among the first to campaign for the abolition of slavery, a campaign that gradually spread across America.

In Birmingham, Joel's brother Benjamin Head Cadbury worked tirelessly for the anti-slavery movement. At least one room in his home was piled high with his voluminous correspondence on the question over the years. After the American Civil War he continued his work for the Society of Freed People of the Southern States, collecting warm winter clothing for women and children, arranging sewing circles and organising the transport of such essentials as bedding and shoes to America. William's uncle, George Senior, also joined the Anti-Slavery Society and contributed funds, along with other family members. With such strong family ties to anti-slavery movements, it was hard for William to understand how they may have been unwitting beneficiaries of a secret slave trade.

In Lisbon, William and Stober hit an impasse. They met ruthless characters such as the Marquis de Valle Flor, the wealthiest São Tomé trader, who was rumoured to subject his slaves to appalling treatment, and to stockpile cocoa to control the price. He was one of many plantation owners who flatly denied having any slaves at all. There were others who refused to speak to William and Stober, or if they did, objected to British enquiries into Portuguese affairs. After all, they reasoned, what about Cecil Rhodes's treatment of Africans in his mines? Or the British Army's mass slaughter of Boer families during the war? How dare the British preach to the Portuguese about morality in their colonies.

William knew that Portuguese cocoa exports from São Tomé to Britain were of only modest significance, accounting for not more than 5 per cent of the total. If British firms stopped buying the island's cocoa in protest, they would lose whatever leverage they held over the Portuguese. Meetings with Portuguese ministers offered some hope. The Minister for Colonies, Manuel Gorjao, conceded that there was an issue, and assured his English visitors that the new Portuguese Labour Decree of 1903 would settle the matter. This decree required workers on São Tomé to be paid a minimum wage, started a modest repatriation fund, and implemented measures to stop illegal recruitment.

Once he was back in England, William Cadbury won the support of all the leading Quaker chocolate manufacturers. Rowntree, Fry and Cadbury were joined by Stollwerck in Cologne in their opposition to slavery, and a debate began as to the best way to tackle the problem. A boycott of São Tomé cocoa, the Rowntrees wrote, 'would mean a serious pecuniary loss to those manufacturers who entered upon it'. Joseph Storrs Fry II replied that this should not stop them 'countenancing a great wrong'. As a first step, the cocoa firms agreed to hire an investigator to travel to the islands and establish the facts. Optimistic that he was making progress in securing a coordinated response from buyers, and that the Portuguese authorities would change their practices, William reported to the Cadbury board that 'things were going to mend'.

Independent of the cocoa firms' investigation, a young English journalist, Henry Nevinson, arrived at the port of Luanda on the Atlantic coast of Angola. He had been hired by *Harper's Monthly Magazine* to make the treacherous journey to the interior of the country, and investigate the rumours of slavery at first hand.

Luanda had once been a key slave-trading centre, but as Nevinson walked through its dusty streets he saw little to hint of it. He took a steamer south to Lobito Bay, another infamous area for the trade, but still he found no direct evidence. Instead, people shrank from his enquiries. He sensed that they were frightened of revealing what they knew, and suspected that anyone who dared to speak out might meet with some mysterious misadventure; poison perhaps, or an apparently random act of violence in the bush.

Despite this, Nevinson and his small party trekked inland, deep into beautiful terrain, 'a land of bare and rugged hills, deeply scarred by weather and full of wild and brilliant colours', wending their way along paths so sheer he compared them to a 'goat-path in the Alps'. They were heading for the heart of Angola's 'Hungry Country'. Still they saw no slave caravans. Did the slavers have advance knowledge of his trip? Could they have changed their route?

But as they continued the 450-mile journey inland, Nevinson began to come across evidence of the trade. Carelessly discarded in the bush, he found the crude wooden shackles that were used to prevent escape. 'I saw several hundred of them,' he recorded, 'scattered up and down the path.' Deeper inland the trail was 'strewn with dead men's bones

. . . the skeletons of slaves who were unable to keep up with the march and so were murdered or left to die'.

Gradually, with discreet enquiries, Nevinson gathered sufficient evidence to conclude that people in the interior were indeed being taken as slaves. Some were sold by their own people. They might be 'charged with witchcraft', he wrote, or be 'wiping out an ancestral debt . . . sold by uncles in poverty . . . or paid as indemnity for village wars'. Local customs made the purchase of slaves easier, he observed, partly because of the 'despotic power of tribal Chiefs', and also 'the peculiar law which gives the possession of the children to the wife's brother . . . who can claim them for the payment of his own debt or the debt of his village'. All too often, however, he found that people were simply seized by agents for the Portuguese in raids on the frontier, or claimed in order to settle extortionate debts owed to colonial authorities.

As Nevinson pieced together the Portuguese labour system he began to expose a cynical pattern of exploitation that left him simmering with rage. Although slavery had officially been abolished, free men were still being made into slaves with the full knowledge and cooperation of the colonial authorities. It was in coastal towns like Benguala that the 'deed of pitiless hypocrisy' that supposedly cleared the Portuguese authorities of wrongdoing took place. Here the slaves seized in the interior were herded in gangs into the tribunal. Paraded before the Portuguese officials, they were asked if 'they go willingly as labourers to São Tomé'. Many were too terrified to speak; those who did were ignored. Official papers were duly completed which apparently 'freed' them, but in fact simply changed their status from slaves to 'voluntary workers' who had agreed of their own free will to toil in the cocoa plantations of São Tomé for five years. This bonded labour was known locally as 'serviçais', but was to all intents and purposes nothing more than slavery.

'The climax of the farce has now been reached,' Nevinson fumed. 'The requirements of legalised slavery have been satisfied. The government has "redeemed" the slaves that its own Agents have so diligently and profitably collected. They went into the Tribunal as slaves, they have come out as contracted labourers. No one in heaven or earth can see the smallest difference.' This 'blackest of crimes' was committed under the full protection of the law.

Although Nevinson took care to conceal the true purpose of his visit, he was convinced he had been poisoned during his stay in Benguala, and was suffering from 'violent pain and frequent collapse'. Nonetheless, he crawled onto a steamer to follow the slaves' journey to São Tomé. He was able to observe them from the upper deck. They were bedecked in 'flashy loincloths to give them a moment's pleasure', a contemptuous token of their new status as 'voluntary' labour. But the *servicais* themselves were not deceived. They knew São Tomé was '*okalunga*' – hell on earth. Nevinson estimated there were 30,000 of these 'voluntary' slaves on São Tomé, and a further 3,000 on Príncipe. Their conditions were harsh, the work unrelenting. On average one adult in five died each year, 'their dead bodies lashed to poles and carried out to be flung away in the forest'.

After his six-month investigation, Nevinson's account was serialised in *Harper's Monthly Magazine* from August 1905. It made for harrowing reading. That same year the Portuguese islands briefly became the world's top cocoa producers, but the cruelty and misery behind the figures was becoming all too evident. This was not, wrote Nevinson, 'the old fashioned export of human beings . . . as a reputable and staple industry'. That, he acknowledged, had disappeared. Nevertheless, 'The whole question of African slavery is still before us,' he concluded. It had merely gone underground: disguised, modified, legalised, but still a loss of liberty. To Nevinson, the enduring horror of it all was 'part of the great contest of capitalism'.

◆◊§❀◊◆

Cadbury and the other Quaker chocolate firms were put on the spot. They had appointed a Quaker investigator, Joseph Burtt, who was due to leave for Africa. Despite Nevinson's revelations in *Harper's*, the young and idealistic Burtt was keen to press on. He was convinced that a second independent report was necessary before they could challenge the Portuguese authorities, as traders in Lisbon were flatly denying Nevinson's account. Was it possible, Burtt suggested, that Nevinson had misinterpreted the situation, perhaps because he spoke no Portuguese, or because of 'the nervous and overwrought condition of his mind'? Burtt thought this might explain why Nevinson feared he was at risk of being poisoned. Such a 'delusion', he said, was common to those 'who have been overdone in Africa'.

It was not long, however, before Burtt too was 'overdone' in Africa. After six months on São Tomé and Príncipe, he fell seriously ill with tropical disease. His fever was so high he could not walk, and he had to be transported in a hammock for part of the hazardous journey across Angola and on to the Transvaal to investigate labour in the South African gold mines. A doctor was sent from England to accompany him, but progress was slow.

In his letters to William Cadbury during 1906, Burtt essentially confirmed Nevinson's findings. The Portuguese administration ran an undercover operation to obtain cheap labour which involved deceit and corruption at every level. 'No imported labourer in São Tomé has ever been returned to the mainland of Africa,' he wrote. 'Children who are born on the estates are the absolute property of the owners.' In the villages in Angola where slaves were seized there was such resignation about the fate of anyone who was taken that his family would 'go through a service of the dead on his behalf'. It was quite clear that the slave trade was every bit as horrific as Nevinson had recounted.

William Cadbury returned to Lisbon to attempt to put pressure on the Portuguese authorities. Once again he was faced with denials, or requests for more time to implement reforms. In response, George Cadbury Senior, Arnold Rowntree and other cocoa directors appealed to the British Foreign Office to confront the Portuguese government directly. On 26 October 1906, William and George Cadbury Senior secured a private meeting with the Foreign Secretary, Sir Edward Grey. They told him that the English cocoa makers were united in their desire to act together on this matter, and were keen to ensure 'that any step we take will be in harmony with any premeditated action of the British Government and be of real use in helping to solve the great African labour problem'. Grey promised to help, and was 'very kind and courteous', George noted. On 6 November, however, George received a letter from the Foreign Office urging discretion. Grey wanted the Cadburys 'to refrain from calling public attention to the question' until he had had a chance to read Burtt's report and 'to speak to the Portuguese minister himself on the subject'. The Cadburys had to contain their impatience.

Unknown to the Cadburys, the Quakers' concern to end slavery immediately was entangled in a far bigger web. Even though the

Foreign Office had promised to put pressure on the Portuguese, there were other British interests at stake that affected negotiations – notably labour problems in the mines in South Africa. The British government was trying to strike a deal with the Portuguese to employ workers from Mozambique in its gold mines in the Transvaal. Consequently, it was in no hurry to antagonise the Portuguese authorities. The Foreign Office was stalling for time.

Burtt returned to England in spring 1907, and reported his findings to the leading cocoa manufacturers. Records show that on 2 May he convinced the Rowntree's board 'beyond all doubt' that the workers on São Tomé were held in 'a condition of practical slavery', and that 'cruel and villainous' methods were used to procure their labour. On 27 June leading directors of the Quaker chocolate firms met to discuss the problem. Around the table were Seebohm Rowntree, the author of *Poverty*, and his cousin Arnold Rowntree; Edward Cadbury, who had just exposed appalling working practices in Britain in *Sweating* and *Women's Work and Wages*, and his cousin William; while the Frys were represented by Roderick and Francis Fry, nephews of Joseph Storrs Fry II. They debated whether the Foreign Office could be trusted to resolve the issue, and whether the cocoa manufacturers should organise a boycott of cocoa from São Tomé. William and others argued against such a move, which they believed would achieve nothing but 'the comfortable assurance that we have wiped our hands of all responsibility in the matter'. At heated meetings on 27 June and again on 4 July, they decided to give the Foreign Office more time, and to try to use their buying power as leverage.

Meanwhile, everything that the Quaker brands of chocolate stood for, the promised land of justice and welfare for all, was being called into question in the British press. Rumours began to circulate that the prosperity of the idyllic Bournville village, with its happy workers producing cheap and delicious chocolate, rested on the unspeakable misery of African slaves. As for George Cadbury, the 'Chocolate Uncle' who created homes for crippled children, was it possible that he was in fact nothing more than a 'serpentine and malevolent cocoa magnate'?

13

The Chocolate Man's Utopia

While the cocoa magnates of England were preoccupied with Quaker idealism and social reform, the slumbering American market was waking up to find it had a chocolate tycoon of its own.

Milton Hershey had returned to the Pennsylvania haunts of his childhood. Exploring the region around Derry Church by horse and wagon he found an isolated farming community still lost in the previous century, lacking the trophies of modernity such as gas and electricity. While his family and critics were astonished at his idea of building on a massive scale in what appeared to be the middle of nowhere, Hershey could only see the area's potential. It was good dairy country – his milk chocolate factory would be surrounded by farms that could supply all the fresh milk he needed.

In 1903 Hershey arranged a survey of 4,000 acres, and bought parcels of land wherever he could, starting with 1,200 acres. A local architect, C. Emlen Urban, was hired to start work on the designs. The factory alone would provide six acres of floor space, filled with the very latest equipment with the capacity to make 100,000 pounds of chocolate each day. It was an extraordinary gamble, but there was no room for caution or uncertainty in Hershey's thinking. His father's advice still echoed in his mind: 'If you want to make money, you have got to do things in a big way.' This would be no mere regional concern, such as his rivals were engaged in. Hershey was going to sell his chocolate across America, from coast to coast. It would be more accessible too: he would sell it at bus stops, railway stations, street vendors and cafés as well as candy shops and grocers. And with mass production, he aimed to lower the price so everyone could afford it. Hershey's would be chocolate for the people.

In 1903 Milton Hershey began building his 'factory in a cornfield'.

Hershey delighted in mulling over the plans with his architect. He knew what he wanted. If money could buy beauty and grandeur, then he would have it. The factory – the humming hub of the enterprise – would look palatial, with elegant limestone buildings set back from the road by a sweeping lawn the size of a football field. It would make Cadbury's little cottage factory look just a little out of date by comparison. Derry Church was to be renamed Hershey, and expanded on a Hershey scale. The ambitious design boasted two main thoroughfares, Chocolate Avenue and Cocoa Avenue, intersected by maple-tree lined streets named after the exotic locations where Hershey bought cocoa: Areba, Trinidad, Caracas, Java, Para, Granada . . . Like Bournville, the charming houses would each have their own gardens, and would be built to differing designs. He planned shops, a library, an athletics ground with a grandstand, and a 150-acre park with an enormous pavilion for skating and dancing. The world would come just to look.

It was a vision that would focus attention on Hershey in a new debate about what the fulfilment of the American dream was all about. This was the age of the gilded millionaires: railroad tycoons like Cornelius Vanderbilt, oil kings like John D. Rockefeller and meat-packing magnates like Gustavius Swift. Newspapers and magazines lapped up stories of their lifestyles, prompting a moral debate: was it right for the fortunate few to possess such staggering wealth while millions faced unimaginable poverty? Was this the embodiment of the American dream, or a grotesque distortion of it?

John Davison Rockefeller conveniently believed that his success reflected God's will. 'God gave me my money,' he is reported to have said. But he was also building a reputation for philanthropy, and in 1901 he founded New York's Rockefeller Institute for Medical Research. Andrew Carnegie, said to be the richest man in the world following the sale of his steel company for $200 million, outlined an entire theory of philanthropy in a series of articles on 'Wealth' for the *North American Review*. He claimed that the rich had a moral obliga-tion to look after the poor.

Ironically, rather than putting an end to the horrific conditions of the workers in his own steel mills, he used his theories to justify them: it was his duty, he said, especially if he was going to give his money away, to make his business as profitable as possible. Carnegie's mills were 'like the gaping mouth of hell', reported *Cosmopolitan* in 1903, their ovens 'emitted a terrible degree of heat', and the workers risked injuries 'of a frightful character'. Carnegie continued to preach what he called 'the Gospel of Wealth', but his views were not always appre-ciated. A theologian at Dartmouth College in New Hampshire, William Jewitt Tucker, was appalled at the patronising idea that millionaires were best-equipped to dispense the wealth they had made from the labour of others. 'I can conceive of no greater mistake, more disastrous in the end to religion if not to society,' he wrote, 'than of trying to make charity do the work of justice.'

Milton Hershey had his own ideas about the fulfilment of the American dream. His staff would be among America's working elite. In the beautiful surroundings of his model town at Hershey, they and their families would thrive, and would be provided with every conceiv-able modern comfort. As the walls of his factory began to rise from the cornfields during 1903 and 1904, the trim figure of Milton Hershey could be seen everywhere, trekking through the mud and dirt. With each day his utopia became more tangible and real.

But the would-be chocolate king was in trouble. Hidden from view in his headquarters for chocolate research, concealed from the curious in a test plant behind his old family home near Derry Church, his struggle to create a milk chocolate recipe was intensifying. As well as a Jersey herd, the facility had its own dairy and milk-processing plant, and a large warehouse filled with kettles and vacuum pans for testing different chocolate mixtures. At night, the grounds were patrolled by guards to ensure there were no intruders. Absolute secrecy was essential to protect the humiliating fact that there was little to guard. Success was proving elusive, but the world must not know.

Like the Swiss before him, Hershey wrestled with the problems of milk processing. Experiments with cream and whole milk failed, so he turned his attention to skimmed milk. The key was to find a way to evaporate the water from the milk in such a way that it could be smoothly mixed with the sugar and cocoa fats. After heating the milk, its flavour almost invariably became poor, burned or scalded. If it managed to condense successfully, it could turn lumpy when it was combined with the chocolate ingredients as the cocoa oils mixed with water in milk. Even if the testers succeeded in making a chocolate bar, it could turn rancid. And then there were the potential pitfalls associated with increasing production. How would they handle the daily input of the milk of 15,000 cows? Could they expand milk processing for mass production? How could they prevent large batches from spoiling?

Hershey was generous but mercurial, kind-hearted but often on a short fuse – shortened still further with worry – making his small staff uneasy. It was common for them to work through the night; impromptu sleeping arrangements were set up in the barn so that an experiment could be watched through the small hours. While their boss's benevolence was beyond doubt, they knew that just one tiny slip-up could be cause for instant dismissal.

There are slightly differing accounts of Hershey's breakthrough. The heroic version has him working late one night with John Schmalbach, a colleague from his caramel factory. After several hours of laborious failures, Schmalbach tried a very slow evaporation of non-fat milk over low heat. He succeeded in reducing the water content of the milk, and added sugar to create a sweet, creamy concoction with no hint of a burned flavour. Better still, they found the mixture could be blended with the ingredients from the cocoa bean without spoiling, to produce a smooth milk chocolate. Hershey was thrilled. In his opinion, the faintly bitter flavour could not be bettered. When the experiment was repeated, it worked again. Slightly sour, distinctly original: the perfect American milk chocolate bar.

According to a more cynical view, the true origins of Hershey's milk chocolate bar are less romantic. Hershey happened to acquire a substantial batch of milk powder from Europe, which had slightly soured by the time it crossed the Atlantic. Reluctant to waste such a large amount, he used it to make chocolate, and found that it sold well. Hershey company officials, however, have always denied that soured milk powder played any part in the breakthrough formula.

Whatever really happened that led to that distinctive Hershey taste, mass production was the next step. With his factory almost complete, Milton Hershey was poised to become the Henry Ford of chocolate. Ironically, the person who would have particularly treasured this moment was missing: Milton's father Henry, the man who had always advocated going for the grand scale, had a heart attack in the unpaved streets of the half-built town. He died in February 1904, just as the elusive American dream was finally being created all around him.

⋘⸖❊⸖⋙

On the other side of the Atlantic, Hershey's prolonged struggles to perfect his milk chocolate recipe were mirrored by those of twenty-six-year-old George Cadbury Junior at Bournville. The Swiss, the Dutch and even Cadbury's old rival Fry were already selling milk chocolate bars. Fry's Five Boys Milk Chocolate, launched in 1902, had very original branding which was having an impact. The wrapper and advertisements featured the facial expressions of a small boy, going from 'Desperation', through 'Pacification', 'Expectation', 'Acclamation' and finally 'Realization' that he is about to eat a Fry's bar. The Rowntrees too, in an unsubtle attempt to cash in on Swiss success, launched Swiss Milk Chocolate, followed in quick succession by Alpine Milk Chocolate and Mountain Milk Chocolate. Although both Rowntree's and Fry's milk chocolates were still based on milk powder, and lacked the quality of the Swiss products, it was surely only a matter of time before they caught up.

Bournville marzipan department, 1900.

The Cadburys had wrestled with the problem of milk chocolate for fifteen years, but had failed to find a breakthrough recipe. The challenge seemed insurmountable: to create and mass-produce a bar that was milkier and creamier than those of the competition. After a research trip to Switzerland, George Junior had tried to release an improved milk chocolate bar in 1902, using condensed milk rather than powdered milk, but it fared no better than the previous attempts. Refusing to give up, George and his team eventually stumbled on a discovery. They came across a milk chocolate bar from Switzerland with a unique taste and texture: rich and creamy, like chocolate velvet. Tests showed that this bar, made on a small scale by the venerable Swiss family firm of Cailler, contained a higher percentage of milk mixed with the chocolate ingredients than other Swiss milk chocolate bars. For George Junior, it pointed the way forward.

George Junior had recently married Edith Woodall, and the young couple settled into a family home called Primrose Hill. Each day, George Junior steered the tiller of his Lanchester to Bournville and walked up the steps to the modest room that passed as the chemists' laboratory. He was joined there by N.P. Booth, the company's first analytical chemist, a confectioner, Otto Unger, an engineer, Louis

Bournville's first laboratory, in which George Junior and his team struggled to create milk chocolate.

Barrow, and a foreman, Harry Palmer. The small team laboured over wooden tables, measuring, titrating and boiling the different mixtures. George was so caught up in the process that it is said that one night, sleepwalking and delirious, he 'rose in the small hours and trundled his young wife, Edith, around the bedroom under the impression that she was a milk churn'.

Late in 1904, an exhausted George Junior and his team hit on the exact combination of temperature, pressure and cooling that would condense the milk in such a way that large volumes of it could be mixed with the cocoa without spoiling. The recipe was indeed creamier and sweeter than Peter's chocolate: rich and very satisfying. Each half-pound bar contained a glass and a half of full cream milk. George Junior thought it good enough to take to the Cadbury board.

Samples were duly handed around. The principal members of the board were George's older brother Edward, his cousins Barrow and William, and his father George Senior. In solemn silence the bite-sized chunks were chewed, pen and paper at the ready to note any flaws with exacting criticism. The seriousness of the occasion was born of many equally hopeful but ultimately unsuccessful trials. Why should this one be any different? Slowly tongue and taste were wrapped around each other. The five gravely sober faces softened, then showed surprise, and then excitement. Another small piece of chocolate confirmed what their tastebuds told them as they licked lips and fingers. This was like Swiss chocolate; no, it was better. It could be launched on the market with confidence.

The date was set for 1905 – the only question was volume. The Swiss were selling thirty tons of milk chocolate a week in Britain. Would the public prefer Cadbury's new product? The board was inclined to be cautious: they wanted to create capacity for producing five tons of George Junior's new chocolate a week. Even this figure seemed optimistic, they reasoned, since it grossly exceeded the consumption of any of their milk chocolate bars to date. George Junior had a different plan. He wanted Bournville geared to twenty tons a week. His idea was to launch in a big way before other firms could copy their recipe. After some debate, he won the backing of the board.

Word of the new and wonderful chocolate spread around the company. The sales team decided to promote the new bar's milk content by calling it Dairy Maid. Just six weeks before launch, however, there was a change of plan. According to the *Bournville Works Magazine*, the idea came by chance from a Plymouth confectionery shop. The daughter of the shop's owner happened to be enjoying a bar of Swiss milk chocolate just as a Cadbury salesman paid a visit. He could not resist telling her about the luxurious new milk chocolate the company was bringing out that would 'sweep the country – Cadbury's Dairy Maid'. The young girl replied, 'Dairy Maid? I wonder you don't call it Dairy Milk, it's a much daintier name.' Six weeks later she received a complimentary slab of milk chocolate with the name she had proposed emblazoned on the wrapping: Cadbury's Dairy Milk.

George Junior's milk chocolate bar finally rolled off the production lines in June 1905. Oddly, it was given hardly any advertising support, and was presented in plain and cheap wrapping to keep the price down. Early sales were slow, and George Junior's prediction of twenty tons a week began to look reckless. The modesty of the launch was not due to Quakerly frugality, but was a result of the company's finances being stretched to the limit by its prolonged struggle with the Dutch over cocoa, the core of its business. Van Houten's alkalised cocoa had cornered 50 per cent of the market in Britain, and sales of Cadbury's Cocoa Essence, its leading brand for thirty years, were falling.

Horse vans were used for local Birmingham deliveries, but in 1906 the sales team also had the first 'Dennis' vans.

George Junior was once again required to come up with a solution – and fast. It wasn't long before he brought a new taste before the board: Cadbury's very own brand of alkalised cocoa. Once again, the board agreed that the new cocoa had a good flavour. The sales teams, however, were baffled as to how to promote this new drink without undermining their old one. The Cadbury name stood for pure, unadulterated products. If they promoted the superior taste and solubility of their alkalised cocoa, they risked drawing attention to weaknesses in Cocoa Essence. Nevertheless, George Junior's new cocoa was launched at Christmas 1906. Packaged in colourful chocolate-brown tins, its name was boldly displayed on the side in large letters: Bournville Cocoa.

In the space of one year, George Junior had launched the two creations that he hoped would save the company from the Swiss and Dutch onslaught: Dairy Milk and Bournville Cocoa. By them he would stand or fall. By them, the company might stand or fall. Like Milton Hershey in America, all he could do was wait.

As the Americans and the British strove to launch their own versions of Swiss milk chocolate, in Vevey, its original creator, Daniel Peter, who had struggled with his invention for so many years, found himself greatly in demand.

Since 1896, Peter had not looked back. Investment in his company, Société des Chocolats au Lait Peter, had increased rapidly. In the three years since 1900, its authorised capital had risen from one million Swiss francs to 1.5 million, and sales reached six million Swiss francs. In addition to his chocolate works at Vevey, Peter developed another plant to meet the rising demand, thirty miles to the north-west, in Orbe.

In January 1904, Peter was approached by another Swiss confectioner, Jean Jacques Kohler, who wanted to join forces. Although Kohler's company was worth less than half of Peter's, its products complemented Peter's range, and it had strong sales teams in France and Switzerland. For sixty-seven-year-old Peter, a merger would enable him to step back from the business, leaving the capable and likeable Kohler in charge. That same month the two firms joined forces to form the Société Générale Suisse de Chocolat, with authorised capital of 2.5 million Swiss francs.

The Nestlé company kept a close eye on the deal. No sooner had Kohler and Peter completed their negotiations than they received a call from Auguste Roussy at Nestlé. Roussy had originally declined to help Peter develop his milk chocolate. Now his company was eager to get into the lucrative milk chocolate market, and had plans to launch brands of its own.

During the summer of 1904 talks took place between Peter and Kohler, and managers at Nestlé. The result was a collaboration that benefited both sides. Nestlé invested a million Swiss francs with Peter and Kohler, and agreed to hand over the manufacture of its own brands of chocolate to them. In return, the Société Générale Suisse de Chocolat was given access to Nestlé's excess milk supplies, its considerable expertise in sales and marketing, and its established position in foreign markets. Effectively the agreement meant that each company would specialise in what it did best: Nestlé would handle distribution and sales, while Peter and Kohler concentrated on quality manufacturing.

But when the deal was announced, the Anglo-Swiss Condensed Milk Company was spurred into action. Founded in 1866 by two American brothers, Charles and George Page, Anglo-Swiss had been

Nestlé's leading competitor for thirty years, and now it wanted to get in on the chocolate bonanza. After failing to secure a deal with other Swiss chocolate firms such as Suchard, it approached Nestlé, and in 1905 the two companies merged to create a Swiss dynamo. The Nestlé and Anglo-Swiss Condensed Milk Company was a truly international concern, with American and European staff and expertise and eighteen factories worldwide: five in Britain, twelve in Europe and one in America.

By skilfully combining their assets and talents, the Swiss had launched a food manufacturing superpower that could already deliver milk chocolate to Europe and America, and had plans to tap into Australia and Asia as well. While the Americans and the British had only recently discovered their recipes, the Swiss deals meant that consumers throughout the developed world would soon be able to sample the delights of quality Swiss milk chocolate.

But even the sophisticated Swiss bankers, with their eyes on the lucrative American market, had not bargained on the well-honed entrepreneurial skills of Milton Hershey. By 1906 the town of Hershey, Pennsylvania, was on the map, and visitors were left in no doubt of that fact. The name proclaimed itself with confidence, writ large and bold and frequently. It was emblazoned on the towering factory chimneys and above the main factory entrance, to remind travellers, should they be in any danger of forgetting, that through his sheer determination in the face of indifferent fate, one Milton Hershey had created a living fairy tale: a town that made chocolate.

The sheer scale of Hershey's operation in America dwarfed the efforts of the Europeans. Apart from his milk chocolate Hershey Bar, he introduced conical chocolate drops called Hershey's Kisses in 1907, followed swiftly by Hershey's Almond Bar. With his leading rivals an ocean away, and the voracious American appetite growing from coast to coast, he had created a chocolate money machine. Sales soared from $1.3 million in 1906 to $5 million in 1911.

Thousands of visitors flocked to see Hershey's miracle, and to marvel at his astounding chocolate factory. The factory in a cornfield was not unlike Bournville, but it was built on an American scale.

Visitors could stroll down the wide boulevards and wander into the centre of town, which boasted landscaped gardens, a zoo, a miniature railroad and a bandstand. Shops sprang up along the main avenues, and there were carefully designed homes for the workers to buy or rent. All the while, Hershey continued to buy up thousands of acres near the town, which he turned into dairy farms to supply the factory. The entire region prospered.

At the very heart of this chocolate utopia was Hershey's own palatial house, High Point. Although modest compared to the homes of other wealthy industrialists of the day, it formed a striking contrast to the plain homestead of Hershey's youth. Approached across a perfect green carpet of lawn, its several wings framed by woodland and a view of the Blue Mountains beyond, it was the embodiment of material success. The white-pillared porch was reminiscent of ancient temples; inside the fine interiors flaunted opulence. Even so, Hershey's Mennonite mother was full of approval for her successful son. She had her own large house on Chocolate Avenue, and enjoyed a respect in the community that was all the more prized after her many long years of struggle. Not content to wallow in her newfound comfort, she combed her grey hair tightly back, donned a starched cap and apron, and continued to wrap sweets for her son as she had always done.

But even Hershey's new wealth could not solve one very personal problem. After ten years of marriage it was becoming clear that his wife Kitty, who was now thirty-eight years old and had long suffered from ill health, was not able to have children. Hershey felt the loss deeply. 'It's a sin for a man to die rich,' he told his friends. He found it hard to accept the doctor's verdict that there was no cure for Kitty's nervous disorder. She hid the growing signs of her weakness and paralysis to spare him pain, and all he could do was watch as his beautiful wife, upon whom he bestowed any jewel she wished for from Tiffany's, deteriorated to the point that she needed a wheelchair.

Kitty began to talk of a new venture, a venture that promised a future: an orphanage for deprived children. Hershey seized on the idea; in the absence of children of their own, this was something they could create together. For both of them it became a lifeline. They could give something real and useful to those in need, the overlooked, the unwanted. According to Joseph Snavely, who worked for Hershey for forty years, they wanted to make themselves 'literally parents to a family of orphan boys'. There would be children about the place, with their innocence and their fresh way of seeing things. The orphans would want for nothing that it was possible for Hershey to provide; they would have all the things he had not had as a child. In November 1909 Kitty and Milton Hershey signed a deed of trust bestowing their orphan school with 486 acres of farmland.

It was a proud moment when the first ten boys enrolled at the Hershey Industrial School. The teaching was informal at first, with the boys gathered around one large table to learn crafts and skills, but they soon moved to a converted barn. The children stayed at local farms, but Milton Hershey loved to visit them every Sunday.

Slowly, the chocolate entrepreneur, the self-made man who lived and breathed the dry air of finance, found the long-neglected role of the father in himself in the company of these desperately deprived and needy children. It was a bittersweet role. He confided to a friend, 'I would give everything I possess if I could call one of these boys my own.'

Milton with boys from the Hershey Industrial School, c.1923.

14

That Monstrous Trade in Flesh and Blood

On 26 September 1908 the London newspaper the *Standard* revealed that Cadbury was profiting from slavery. The white hands of the Bournville chocolate-makers, it said, 'are helped by other unseen hands some thousands of miles away, black and brown hands, toiling in plantations, or hauling loads through swamp and forest'. Their wretched lives formed a stark contrast to those of the young women employed at Bournville, who 'visit the swimming bath weekly and have prayers every morning'.

The *Standard* deplored what it clearly regarded as hypocrisy. 'It is that monstrous trade in human flesh and blood against which the Quaker and Radical ancestors of Mr Cadbury thundered in the better days of England,' claimed the editorial. 'And the worst of all this slavery and slave driving and slave dealing is . . . to provide a sufficient number of hands to grow and pick the cocoa on . . . the very islands which feed the mills and presses of Bournville.'

At the next board meeting at Bournville, the mood was sombre. William felt traduced; his years of effort to put an end to this evil practice had been ignored, and his planned visit to Angola mocked. The article created the impression that the Cadbury family had deliberately delayed resolving the issue of slavery in order to continue making a profit from São Tomé cocoa beans for as long as possible, all the while trumpeting its virtuous practices at home.

Those present at the meeting knew that this simply was not true. There were, however, questions to be answered. Why had it taken William so long to investigate conditions on São Tomé? Had he been too trusting, especially of politicians like the Foreign Secretary Sir

Edward Grey, who appeared to have promised much and done nothing?

George Senior stood by his nephew. He knew the Cadbury board had agreed that its aim was 'to put a stop to the conditions of slavery – not merely to wash our own hands of any connection with them'. If they had stopped buying cocoa beans from São Tomé, they would have lost what little leverage they had over the Portuguese to effect reform. Yet in the eyes of the *Standard* and its readers, they looked hypocritical and lacking in integrity. As head of the firm, George Senior was directly accused. His staunch Quaker principles, his wealth and the utopian village he had created appeared to rest on the back of slaves. It was a shaming indictment for a man of God. The board decided to sue the *Standard* for libel.

In October, William embarked on a trip to West Africa with Joseph Burtt. Their aim was to find out if there had been any changes following the years of diplomacy with the Portuguese and British Foreign Office. But the Portuguese authorities seemed to have learned of their visit in advance. They made sure there was nothing to be seen, and assured their English visitors that signing up new *servicais* had been 'suspended'.

William and Burtt were not convinced. Although they did not catch the Portuguese red-handed, they were suspicious that new slaves were being smuggled onto São Tomé at night. They saw known slave traders with Portuguese officials when they visited Angola, and the ill-concealed antagonism shown to them by the Portuguese governor general failed to reassure them that the colonial authorities were genuinely implementing reform. The underground slave trade, they concluded, was being hidden more skilfully than ever. They returned to England in March 1909, and William discussed their findings with the managements of Cadbury, Fry and Rowntree. A week later the British Quaker firms announced that they were boycotting cocoa beans from São Tomé and Príncipe.

Several European companies followed their lead, but as George Senior and William had feared, it was soon apparent that a boycott would not stop the slave trade. German and American brokers simply stepped in to purchase the beans from São Tomé, which were superior to any grown elsewhere in Africa. Working through the Anti-Slavery Society, Cadbury sent experts to Germany and America to alert foreign cocoa companies to the situation.

Cadbury and the other Quaker manufacturers were keen to investigate alternative cocoa plantations in Africa. The first cocoa seeds
had reached the Gold Coast – now known as Ghana – as recently as
1879. Although the climate was suitable, the beans from this region
were of poor quality, and on his recent trip to Africa William had
bought a small site at Mangoase in the Gold Coast to research ways
of improving production. He soon found that local dealers paid farmers
the same low price for cocoa beans regardless of their quality. Bags
would be filled with damaged or wet beans, which rotted during ship-
ment and ruined the whole batch. Working with local producers,
Cadbury's and Fry's representatives made it clear that they were
prepared to pay the market price for quality cocoa. They taught local
farmers how to carefully dry and ferment the cocoa, and production
from the region soared.

Meanwhile, on 29 November 1909, the case of *Cadbury Bros. Ltd.
v. the Standard Newspaper Ltd.* began. The courtroom in the imposing
Victoria Law Courts in Birmingham was packed with press, family
members and curious members of the public. Lining up to represent
the two sides were the most renowned lawyers of their day. Sir Edward
Carson, the Conservative MP and barrister who was appearing for
the *Standard*, was known for his theatrical style and a flair for inter-
rogation that had earned him a reputation as 'the best cross questioner
in England'. He had become well known during Oscar Wilde's libel
trial in 1895, when he had elicited the evidence that revealed Wilde's
homosexuality. Magazines such as *Vanity Fair* printed caricatures of
the tall, spare Carson, his chin thrust forward and his bushy eyebrows
arched as he posed a characteristically contemptuous question.
Cadbury had hired the acclaimed Liberal barrister Rufus Isaacs. Sober
and exacting, he was among the very few who could take on the
celebrated Carson.

The case had come to symbolise more than libel. For many it
represented a titanic clash over the ethics of the Liberal and
Conservative parties. The battle lines were neatly described by
Carson's biographer, Edward Marjoribanks: 'The greatest Conservative
leader of the day was engaged for a Conservative organ, against the
most brilliant advocate in the Liberal Party who was in turn attacking
on behalf of the honour of a great Liberal family not unconnected
with the most powerful Liberal organ [the *Daily News*].' The Tory

Standard was edited by Howell Glynne, a keen supporter of Cecil Rhodes, Joseph Chamberlain and the Boer War. George Cadbury Senior's Liberal *Daily News* had not only passionately opposed the war, but had also exposed abuses of British power in South Africa, notably the cruel exploitation of Chinese coolie labour in the mines. This, said the *Daily News*, was slavery by the back door. Yet now Cadbury's paper and the Liberal establishment were under scrutiny for apparently turning a blind eye to slavery in the São Tomé cocoa plantations.

William Cadbury was the first to take the stand. He was now forty-two, his dark hair thinning, his manner quiet and unassuming, his Quaker convictions imbuing him with a sense of composure. During four days of questioning the great barrister made him spell out to the court every detail of his investigation.

> *Edward Carson*: Men, women and children, freely bought and sold?
>
> *William Cadbury*: I don't believe, as far as I know, that there is anything that corresponded to the slave market of 50 years ago. It is now done by subtle trickery and arrangements of that kind. I don't think it is fair to say that in Angola there is an open slave market.
>
> *Edward Carson*: You don't suggest that because it is done by subtle trickery that makes it better?
>
> *William Cadbury*: I do not.
>
> *Edward Carson*: But in some cases they are bought and sold?
>
> *William Cadbury*: In some cases they are bought and sold.

When Rufus Isaacs took over the questioning, William had a chance to describe the steps he had taken since he first became aware of the slavery problem in 1901. Cadbury had tried to use its buying power to bring about real reform, he explained. The company had appointed its own investigator, Joseph Burtt, to present the Portuguese authorities with definitive proof of slavery, rather than unsubstantiated allegations that could be easily dismissed. William described his repeated trips to Lisbon to pressurise the Portuguese authorities, and his discussions with the Foreign Secretary. 'We went on buying the cocoa,' he explained, 'because we were absolutely advised by the highest authorities we could consult that it was the best thing we could do in the interest of reform of labour conditions in West Africa.'

George Cadbury Senior was the next to be called to the witness stand. At seventy, he was still a tall man, with a commanding physical presence, but he walked more slowly than in his youth. His wife Elsie, watching anxiously from the gallery, was aware that he was feeling the pressure. It was hard for her, knowing her husband's passionate commitment to just causes, to watch as he was subjected to a humiliating cross-examination.

George explained that the company was trying to bring about change, but he seemed frail, and did not speak with his nephew's self-assurance. Carson pressed on. As the owner of the *Daily News* and 'the great champion of the conditions of labour of South Africa and the Congo . . . why then did Cadburys – those perfect gentlemen . . . pay one million [and] three hundred pounds . . . to the slave dealers in Portugal?' In the close atmosphere of the court, George Senior sometimes failed to answer Carson's questions effectively. He relied on honesty, which was soon made to look foolish or worse at the hands of the clever Carson. There were cries of 'Shame!' in the courtroom. To those sympathetic to the *Standard*, the acclaimed philanthropist appeared reduced to a pitiable old man.

To everyone's surprise, Carson did not call any witnesses to support the *Standard*'s accusations. Justice Pickford summed up the issues for the jury. They 'were not there to decide whether Messrs Cadbury took the right course, or whether any better course could have been taken', he declared. There was only one question the jury had to resolve: 'Were [the Cadburys] for the purpose of preventing an attack on their character as philanthropists, and at the same time to postpone any final decision as to not purchasing cocoa grown by slave labour, were they purporting to take steps which were of an ineffective nature in order to obtain a mitigation of the evils?' In effect, had the chocolate firm been operating a dishonest plot to enable it to profit from slave-grown cocoa?

The jury returned within the hour. The foreman rose. The jurors found in favour of Cadbury Brothers, he said. But when he announced the damages against the *Standard*, the result was unexpected.

'One farthing!'

The outcome was enough for George Cadbury Senior. 'It has been a long ordeal,' he wrote later to his son Henry. 'Mother was there most of the time and I think is thoroughly tired out . . . How thankful we are for the result.' He felt relieved that the 'slanders hurled against us had been cleared'. He was also spared legal fees, since the *Standard* was ordered to pay the costs. For William Cadbury it was not so easy to come to terms with the result. Although the jurors had agreed that Cadbury had been libelled, the contemptuous damages they awarded implied that they were not impressed by the company's handling of the slavery issue, notably the long delay before it instituted a boycott of São Tomé cocoa beans.

There was one benefit from the lawsuit, however, at least in the short term. News of the case reached America, where it prompted chocolate manufacturers to join the fight to urge the Portuguese to end slavery. Caught in the glare of international publicity, Portugal finally halted the transport of slaves from Angola in 1909, and according to one account, 14,000 slave workers were repatriated from the islands. In the longer term, however, the issue was far from resolved. The British Anti-Slavery Society continued to receive reports of abuses, and the Foreign Office faced repeated criticism for its failure to persuade the Portuguese to improve conditions for workers on the islands.

In time, the new agricultural techniques introduced to the Gold Coast by Cadbury and Fry brought improvements in production. In 1910 the Gold Coast farmers harvested 26,000 tons of cocoa, more than the notorious islands of São Tomé and Príncipe. The following year this jumped to 40,000 tons – a world-record national harvest. The 'food of the gods' from the New World was becoming established as an Old World crop. *Theobroma cacao* made its way west along the high grassland savannahs of the Ivory Coast and Sierra Leone, east into Nigeria and the Cameroons and south into the great Congo Basin, paving the way for Africa to become the world's leading producer of cocoa.

With demand for chocolate soaring in the West, cocoa manufac- turers were urgently seeking new sources of beans. Dutch colonialists had already taken seedlings to Java and Sumatra, and the British were establishing plantations in Ceylon and India. The cocoa tree, its seeds carried by missionaries and colonists, even graced the shady beaches of far-flung Pacific islands such as Vanuatu and Samoa. The exotic plant, once a precious currency of ancient civilisations in the Americas, was spreading around the world, following the shadowy blue fringes of the great equatorial rainforests

George Cadbury Senior had a stubborn streak. Although there were some in the family who believed that the *Daily News*'s own investigations into abuses had goaded the Tory press into making its libellous accusations, George Senior regarded his purchase of the newspaper as money well spent. Since the launch in 1906 of HMS *Dreadnought*, the new terror weapon of the day, the naval arms race between Britain and Germany had accelerated. George saw it as his duty not just to champion liberal reform at home but also the cause of diplomacy and peace overseas.

Shortly after the court case George Senior learned that two Liberal papers, the *Morning Leader* and the *Star*, with a combined circulation of half a million, were up for sale. He did not want to see them fall into Conservative hands, later telling his son Laurence, 'It seemed a very serious matter to let them go into the hands of those who might seek to promote war with Germany and would oppose measures of social reform.' George approached the Rowntrees to discuss the possibility of the two Quaker families buying the papers together. Joseph Rowntree's nephew Arnold, who already had experience running the *Northern Echo* and the *Nation* for the Rowntree Social Services Trust, was keen to work with George Senior.

After they had bought the papers, however, they faced a serious dilemma. Both the *Morning Leader* and the *Star* published gambling news, which provoked condemnation from their Quaker friends. The views of 'Old Joe or 'Captain Coe' on the day's horseracing had no place in a Quaker-run newspaper. Joseph Storrs Fry II's brother, the successful barrister Sir Edward Fry, led the charge of hypocrisy. The very men from the Cadbury and Rowntree families who supported the National Anti-Gambling League, he declared, were encouraging gambling in their own newspapers.

The two families, however, had already tried removing the betting news from the *Daily News* and the *Northern Echo*, and in both cases it had led to a catastrophic fall in circulation. There was a real possibility that if it was removed from penny and halfpenny papers like the *Morning Leader* and the *Star*, the papers might go under, or have to be sold to an owner who didn't share Quaker ideals. 'It was evident that the *Star* with betting news and pleading for social reform and for peace,' George Senior explained to Laurence, 'was far better than the *Star* with betting and opposing social reform and stirring up strife with neighbouring nations.'

It was not long before the Tory press picked up news of the split in Quaker thinking. Journals such as the *Spectator* took up arms against the apparently ever-expanding 'Cocoa Press'. Quaker virtue was once again on the line. George Senior was 'held up to execration as an

George Cadbury, seen here with a copy of the *Daily News*, was accused of taking 'the devil into partnership to aid the Almighty'.

odious example of the sleek hypocrite who profited by the degradation and vice of others'. He was particularly troubled by the continued accusations of fellow Quakers such as Sir Edward Fry, who implied that he had taken 'the devil into partnership to aid the Almighty'. It reached the point, declared the *Manchester Guardian* in an editorial, that the Cadburys and Rowntrees were being 'assailed with such severity' that anyone might suppose 'that they had introduced a gambling newspaper for the first time into the white robed company of the London daily press'.

On 14 November 1911, with a view to streamlining the organisation of the papers and withdrawing from their management himself, George Senior set up the Daily News Trust. In transferring his ownership of the papers he articulated how he hoped the trust would be run. The document is a testament to his desire to apply Quaker beliefs and Christian thinking to the worlds of business and journalism:

> *I desire in forming the Daily News Trust that it may be of service in bringing the ethical teaching of Jesus Christ to bear on National questions . . . for example that Arbitration should take the place of War and that the spirit of the Sermon on the Mount . . . should take the place of Imperialism and of the military spirit which is contrary to Christ's teaching that Love is the badge by which a Christian should be known. The parable of the Good Samaritan teaches human brotherhood, and that God has made of one blood all nations of men. Disobedience to this teaching has brought condign punishment on nations; and though wars of aggression have brought honour and wealth to a few, they have always in the long run brought suffering upon the great majority . . .*

Throughout his life, the New Testament served as George Cadbury's guiding principle. For him, it was simple: if everyone followed the teaching of Christ, people and nations could live in peace together. His faith anchored his thinking and unified all spheres of his life. It speaks volumes for the times that the editors of his newspapers endeavoured to support his view.

Ironically, the political will was already in place that would make such a close marriage of religion and business less urgent. That very year a new law was enacted that brought fundamental reform to at least one social issue that had troubled the Quaker conscience for generations: poverty. The impetus for change had come in part from the Quaker community itself, through influential writers such as Seebohm Rowntree.

After the publication of *Poverty* in 1901, Seebohm had toured the country promoting a new vision of social responsibility. His research inspired prominent politicians, such as Lloyd George, the Chancellor from 1908, and Winston Churchill, who had now crossed the floor of the House to become a Liberal MP. 'This festering life at home,' Churchill wrote of the British poor, 'makes world-wide power a mockery, and defaces the image of God upon earth.'

Such influential thinking coincided with another forceful voice coming from the workers themselves. Still labouring under the harsh regimes of the previous century, the vast majority felt alienated from the rest of society. Across Britain, workers were waking up to their own power. Some enlightened employers, such as the Rowntrees in 1906, had voluntarily set up their own pension schemes and provided benefits including free medical care and education, but they now found their workers joining unions and making more demands.

For Joseph Rowntree, who had put up £10,000 of his own money to start the firm's pension fund, it was an unsettling time, reflecting an uneasy shift in the relationship between an employer and his staff. But his sons joined the growing number of employers who believed union membership was the way forward. Across Britain, the Labour Party was gaining support, and membership of the new trade unions reached 2.6 million in 1910. All this reflected a widespread recognition that the Victorian toleration of poverty had no place in twentieth-century Britain.

This momentum for change led to a series of groundbreaking Liberal reforms. After vigorously lobbying through the *Daily News*, the National Old Age Pensions League and the Anti-Sweating League, it was rewarding for George Senior and Edward to see that change was finally possible. The Old Age Pension Act of 1908 provided a means-tested income of between one and five shillings per week for people over seventy. It was collected at the post office, to avoid any of the stigma associated with the old parish payments. The majority of those eligible in the first year were women with incomes of less than £31.10s a year. The Trade Boards Act of 1909 created boards that could set minimum wages in trades that were notorious for sweatshop labour, such as tailoring. That same year saw Lloyd George's revolutionary 'People's budget', which set out to bring about a redistribution of wealth, with higher rates of tax for those earning above £2,000 to fund further reform. Following the National Insurance Act of 1911, the state provided a basic level of unemployment and sickness benefits. At last, the law of the land recognised the plight of those overtaken by the struggle of life. By today's standards the amounts they received were modest, but the ambition for reform was huge.

In time these sweeping Liberal reforms would lead to the abolition of the Poor Laws, with all the shame of miserly welfare dispensed by the parish under laws that had their origins in Tudor times. Workhouses, which many felt punished people for their poverty, survived until 1930; even after that, many remained – renamed as Public Assistance Institutions – until 1948. But much of the grinding poverty caused by wages on which a family could not possibly survive was relegated to the past.

One unintended effect of the new legislation was to help to distance religion from business. The need for charitable Quaker businessmen like Joseph Rowntree to provide for their workers out of their own pockets became less urgent. The reforms helped to forge a framework for modern social welfare, meaning that there was less need for random, and to some patronising, individual acts of charity. What had formerly been the domain of men of God was becoming the official business of the state.

◆◊✿◊◆

Hostility to the Cadburys' and Rowntrees' 'Cocoa Press' may have been exacerbated by the resounding success of the chocolate firms in the early part of the twentieth century. Far from being damaged by the *Standard* libel case, Cadbury was becoming an increasing threat, even to its Quaker friends and rivals at Fry in Bristol. In 1905 Cadbury's sales of £1,354,948 were only just behind Fry at £1,366,192, while Rowntree had sales of £903,991. For Cadbury, overtaking its long-time ally and rival became a realistic goal.

Cadbury at last had a bar that was a strong contender in the chocolate wars. The titanic struggle between Swiss and British producers was being fought in confectioners' and grocers' shops up and down the country. The weapons were chocolate confections with irresistible names, backed by travellers in motor cars, poster campaigns, price wars and publicity stunts.

Swiss chocolate was still so prized that at first George Cadbury Junior's Dairy Milk had trouble selling even two tons per week. But word of its quality spread, and by 1910 it was the clear favourite among British chocolate brands, and was well on its way to becoming Cadbury's bestselling line as chocolate sales began to catch up with those of cocoa.

It was a wonderful vindication for George Junior as sales finally approached his initial ambition of twenty tons a week. Dairy Milk proved so popular that Cadbury introduced it to other lines such as the Fancy Box, and as a coating on Easter eggs. George Junior supervised the creation of a dedicated milk-condensing plant at Knighton in Staffordshire to satisfy demand. His Bournville Cocoa was also a great success, overtaking the long-established Cocoa Essence by 1911.

Transporting milk to the Knighton factory in Staffordshire.

Cadbury's export trade was also growing rapidly under the meticulous supervision of George Junior's older brother Edward. According to his colleagues, Edward operated 'at high pressure, made swift decisions and was not always easy to work with'. Yet there was no doubting his business acumen. At the time when he took over the export department, entire continents were managed by a single traveller. In just a few years, after extensive travels around the globe, Edward established sales forces in China, South America, Canada and the West Indies. Shrewdly exploiting the commercial potential offered by the British Empire, Edward's efforts meant that exports soon made up 40 per cent of Cadbury's total sales.

After a pursuit that had lasted half a century, Cadbury finally overtook Fry in 1910 to become Britain's largest manufacturer of cocoa and chocolate. That year Cadbury's sales reached £1,670,221, compared to Fry's £1,642,715. Rowntree was also becoming a threat to Fry with its sales having climbed to £1,200,598. The efficiencies gained by the company in moving to Haxby Road, combined with the continued success of its gums, pastilles and Cocoa Elect, had brought about a period of sustained growth.

Despite being afflicted with deteriorating eyesight, in Bristol the eighty-four-year-old Joseph Storrs Fry II continued to hold onto his power as company chairman. His nephews and cousins struggled to overcome his aversion to change, but he failed to inspire his development team to come up with a satisfactory competitor to Dairy Milk. Ignoring the evident rewards that had come to both the Cadburys and the Rowntrees by their moves to purpose-built modern sites outside Birmingham and York, he repeatedly opposed any move from the centre of Bristol.

For him, the awkward conglomeration of premises the firm occupied was sacrosanct, hallowed by centuries of Quaker tradition, 'beautiful' in his eyes. But those eyes, always attuned to the inner vision of his youth when those in charge at Fry had 'waited upon the Lord', were growing dimmer. New buildings, new products – the modern generation was obsessed with change. They missed the point: 'Unless the Lord build the house, they labour in vain that build it.' The world of chocolate-making faded as he grew older, and his religious interests became more focused. At last, his sight finally gone, his world lit by faith alone, Joseph died, aged eighty-seven, on 7 July 1913. He had always remained a man of God first, a chocolatier second.

Although Cadbury had overtaken its chief British rival, the Swiss remained beyond reach. The merger of Peter-Kohler in 1904 and its marketing agreement with Nestlé had precipitated considerable growth. Swiss interests converged further when in 1911 the family firm of Cailler approached Peter-Kohler and another deal was signed. Nestlé held a 39 per cent stake in the united firm of Peter-Cailler-Kohler, and the Swiss giant that emerged was a colossus, apparently unbeatable. More than half of all chocolate consumed worldwide was Swiss.

But the bitterly fought battle among the European chocolate

empires was about to be eclipsed by a far more serious conflict. On 28 June 1914 an assassin's bullet in Sarajevo triggered the chain of events that would see the boundaries of Europe, and the commercial empires within them, totally redrawn.

15

God Could Have Created Us Sinless

In the years before the First World War, Barrow Cadbury had joined Quaker and other Church leaders in a high-profile peace campaign. He was among those presented to Kaiser Wilhelm II at the Royal Palace at Potsdam in 1909 to press the moral case for peace. Two years later, German churchmen were introduced to the King and Queen at Buckingham Palace. A second British deputation went to see the Kaiser in 1913, and made plans for a peace conference at Constance, on the Swiss–German border, in early August 1914.

Such pacifist idealism may seem otherworldly, even naïve, in the face of the mighty German and British military industrial machines. But since its earliest days pacifism had been central to the Quaker movement. But by July 1914 the impetus for war was irresistible. On 1 August, Germany declared war on Russia. Two days later, Germany was at war with France. On 4 August, Germany invaded Belgium, and Britain declared war on Germany. As the religious leaders were making their way to Constance, Europe's armies were on the move. The peace conference was forced to disband.

In Britain, Lord Kitchener, who as a Boer War commander had been responsible for the policy of herding women and children into concentration camps, was now Secretary of State for War, and appeared on the famous poster urging 'Your Country Needs You'. Hundreds at the garden factory in Bournville answered his call.

At the *Daily News*, George Cadbury Senior faced a crisis of conscience. Unlike the Boer War, which he viewed as plainly unjust, this conflict brought him 'face to face with the fundamental tenets of religious persuasion'. As he paced the large wood-panelled rooms of the manor house, a set of Quaker rules devised in the English shires during the seventeenth century must have seemed remote from the present situation.

The war split the Quaker community in two. For some there were no circumstances under which a Quaker could deviate from the doctrine of non-violence. For others, such as George Senior, this was not a simple issue. Although he acknowledged the complexity of the causes of the war, and the faults on the British side, he believed that the prime reason for this world calamity was the aggression of Germany. While he remained 'a very strong believer in Peace', he wrote that German militarism 'must be crushed'.

The issue was made more urgent for George by the fact that his and Elsie's three sons had recently reached adulthood. Despite the long-standing family commitment to pacifism, their headstrong youngest son, twenty-one-year-old Egbert, known as Bertie, left the Quaker movement immediately to sign up as an Able Seaman with the navy. He and his friends 'were all in a desperate hurry to enlist because we thought the war would be over by Christmas and unless we were quick we might miss all the fun'. Bertie's brother, twenty-four-year-old Norman, had trained as an engineer and was working for the Electrical Mechanical Brake Company in West Bromwich. Soon he was swept up in a most un-Quakerly activity, manufacturing bomb parts, shells and artillery hubs, and later in the war gears and track links for tanks.

George and Elsie's oldest son, twenty-five-year-old Laurence, was keen to volunteer for the army, but remained committed to the Quaker movement. 'I strongly urged – as I am a well known advocate of Peace,' George Senior said later, 'that Laurence should join the Ambulance Section.' On 21 August 1914 Arnold Rowntree and others launched an appeal in the Quaker magazine the *Friend* for volunteers to form a Friends Ambulance Service. Laurence was one of the forty-three volunteers who left for the battle front with the first unit.

When they reached France, they learned that wounded from the Battle of Yser in south-west Belgium were in need of help. According to the Friends Ambulance Unit records, when Laurence and his Quaker friends reached the sheds of Dunkirk railway station, where the men were being housed, 'a terrible sight met their eyes'. Light filtered through the windows revealing thousands of men with

Laurence Cadbury.

appalling injuries, lying where they had been left. Many had been there for three days and nights, 'practically untended, mostly even unfed, the living, the dying, and the dead side by side, long rows of figures in every attitude of slow suffering or acute pain, of utter fatigue or dulled apathy, of appeal or despair'. The air was thick with the sickening smell of gangrene. Working day and night, the Quaker volunteers treated the wounded and organised transport for 6,000 men to be transferred to hospital.

In early November Laurence's convoy was on the road again. The next stop was an ancient town in west Belgium caught under intense bombardment: Ypres.

❧✿❧

During the autumn of 1914, as Europe faced the horror of warfare in the industrial age, Milton Hershey's forty-three-year-old wife Kitty was also struggling to survive. With each passing day, Kitty seemed a little more diminished; her smile fragile, her breathing difficult. Swaddled in the most expensive furs, she was still cold; pampered with every conceivable diversion, she did not have the strength to lift a book. Slowly her mysterious nervous ailment robbed her of her last vestiges of dignity; despite her husband's devotion, she was utterly helpless. On a trip to Philadelphia in March 1915, Milton left her side to get some champagne to cheer her. When he returned, he was too late. Kitty had died.

Milton was utterly distraught. Here was a man who apparently could do anything, who had built a nationwide brand, a multi-million-dollar business and a celebrated town, but his massive wealth could not save Kitty. With his wife lying in a bronze casket, fresh flowers ordered to decorate her vault for years into the future, he threw himself into what he knew best.

Work was Milton's world. In the months after Kitty's death he cut a mercurial, frenetic figure, with an occasional unpredictable streak that made employees worry about getting on his wrong side. At times complete strangers might be the recipients of his unexpected good will, such as a cripple he spotted from his car who suddenly found himself with sufficient funds to pay for the surgery he needed. At other times, a committed and long-serving employee might find himself in the firing line over something that was not his fault.

Hershey's Chocolate Factory continued to mint money, with sales approaching $10 million a year by 1915. Each week four hundred vehicles queued to load up with its goods. Hershey's influence over the district was spreading as his farms extended over almost 9,000 acres. Visitors arrived in their thousands to see the wonderful sights of the famous chocolate town.

In early 1916 Milton Hershey paid a visit to the Caribbean island of Cuba. Cuba had recently been liberated from Spanish control, and the USA was flexing its own imperial muscle, seeking colonies of its own and bringing American style and commerce to the island. Hershey seized the opportunity. With the war posing a threat to Atlantic shipping, he feared a sugar shortage might affect his chocolate money machine. As he settled into a luxurious hotel in Havana enjoying views over Havana Bay, he wondered if there was a way he could control his own sugar production.

What began as a modest trip through the sugar plantations with a local guide culminated in a gamble on a typically Hershey scale. Milton toured the northern coast of Cuba with the boundless enthusiasm of a young man. Tirelessly he trekked through fields of sugar cane, not noticing the heat, viewing the vigour of the canes, the flowing water, the rich earth of the glorious, fecund island. Hershey was in a buying mood: 10,000 acres thirty-five miles east of Havana were a start. It was like beginning anew, and required all his energy; and he had energy to squander as he tried to blot out the memory of Kitty's loss.

Hershey raised the funds for his venture by selling securities and creating a new company, the Hershey Corporation. This put him in a position to proceed simultaneously with the construction of a giant sugar mill, a grand hotel and, of course, another model town, Hershey Cuba, with over 180 cottages, each with its own garden. He wanted his workers to have the best of the latest technology. Their houses would boast such extravagances as taps with running water and electric lights. He continued to buy land on Cuba – eventually totalling some 65,000 acres – and made plans for an electric railway that would cross it from Havana to Matanzas: the Hershey Cuban Railway. While eyebrows were raised back in Hershey, Pennsylvania, at the extravagance of the venture, Milton felt invincible as sugar prices continued to climb.

He became a familiar figure in Havana, the very image of wealthy benevolence as he strolled through the tropical landscape in his white suit and panama hat. His mother sometimes accompanied him on his trips there, when they stayed in old-world haciendas and enjoyed the view of Havana Bay. Milton continued to try to spend his way out of his grief, and it seems that the expatriate society in Cuba afforded him some relief. Millionaires he had long admired, such as Henry Ford, also relished the privacy of the island. These were Havana's boom years, when exotic luxuries were always available to the super rich, and the tropical air was fragrant with the scent of money.

Hershey's business continued to expand, and he could now control the price of one of his main raw materials. He almost seemed blessed with the Midas touch. But as he drank champagne, smoked fine cigars and gambled in Havana's nightclubs, the widowed, childless Milton had no one to whom he could bequeath his vast wealth, and all the while his chocolate-making money machine grew daily.

On 31 May 1916 the mightiest fleets the world had ever seen, each equipped with an array of weapons sufficient to blow its opponents out of the water, confronted each other in the North Sea near Jutland in Denmark. With a horrible inevitability, soon all was chaos, the dark shapes of the dreadnoughts and battle cruisers barely visible in the smoke of battle as they spilled out thunderous gunfire. The noise was deafening, the smoke alternately obscuring and revealing scenes of horrific destruction as shells ripped into man and machine. Fourteen British and eleven German ships were lost in the Battle of Jutland, and 8,500 young men gave their lives. A month later, one of the bloodiest military operations ever recorded unfolded at the Somme in France. In one fateful day, 1 July, 19,240 British men died and more than 35,500 were wounded – the greatest losses in a single day in the entire history of the British Army.

News of the battlefields of France and Flanders reached George Cadbury Senior and his community at Bournville not only through the press, but in letters from Laurence and others in the Quaker convoys. For George Senior, the war was proving to be a considerable test of faith.

Friends Ambulance Unit ambulance train.

How could anyone make spiritual sense of the depths of suffering that new technology was leaving in its wake? The poisonous yellow-green plumes of chlorine gas floated like clouds over the trenches at Ypres, burning eyes, dissolving lungs and causing a slow and terrible death by asphyxiation. Fighter planes, the new terror of the skies, were armed with machine guns and rained bombs on those below. Then there were the tanks that crushed men into the mud beneath their treads, and the young men torn to shreds in the barbed wire of no man's land. All this evil handed out so impersonally demanded an explanation. The industrial age, which had brought so many benefits, had also unleashed a monster: killing on a scale never before seen in human history. The world order, it seemed, was collapsing. This orgy of mindless killing was incomprehensible.

George Cadbury felt a pastoral concern for his flock at Bournville. Faith was at a crisis point as many wondered how God could allow such a war. Although George too had unresolved concerns arising from the conflict, he felt it was his duty to visit the families who attended the Bournville Meetings. They wanted to discuss their worries and fears. They wanted reassurance, and an answer to the question of how God could permit such evil.

George gave no sign that he was wrestling with the same thoughts himself, but spoke only of faith: 'If there were no sin, there could be no free agency. God could have created us sinless; then there would have been no freedom of choice and no test of our obedience and love for him.' Then he would ask his questioners if they had ever been tempted by evil, but resisted that temptation. Drawing on his experiences in the Adult Class, he said that he had seen many times that 'yielding to sin brought no real or lasting happiness, but resisting temptation did'. He quoted from the Bible: 'Blessed is the man that endureth temptation for when he is tried he will receive the crown of life, which the Lord has promised to them that love Him.'

It is hard to gauge the impact of these visits, or to know if they were appreciated. George Cadbury had a patriarchal relationship with the inhabitants of Bournville. Doubtless they were in awe of their employer and benefactor; some may have been moved and helped by the strength of his faith.

But in the country as a whole there was growing hostility to pacifists. The Cadbury family made no secret of their views, and in

November 1916 a military inspector was ordered to examine George's activities to find out if he was funding anti-war or anti-conscription campaigners. The inspector even insisted on going through George's personal accounts. George had nothing to hide: three-quarters of his income was given away to philanthropic causes, but none were anti-war campaigns. His family, too, was committed to the war effort.

Twenty-year-old Molly, his youngest daughter, had followed her older brother Laurence to France soon after the outbreak of war to work in a military hospital. Laurence was building a reputation in the Friends Ambulance Unit, with eighty vehicles under his control. Their sister, twenty-two-year-old Dolly, worked as a nurse at Fircroft Working Men's College in Selly Oak, which had been converted into a hospital for wounded servicemen, as had The Beeches.

George's youngest son, Bertie, had initially been posted to Great Yarmouth, from where he took part in mine-sweeping operations in the North Sea. 'Don't tell Mother or Father,' he wrote to Laurence after a particularly reckless trip when his unit had unknowingly sailed straight through a minefield, 'or they will have hell's own needles.' Nine months later he transferred to the Royal Naval Air Service. His task was to attack the German Zeppelins which had brought a new terror to the towns of Britain. At first the British were hopelessly ill-equipped. The Zeppelins flew on dark nights, but there were only two searchlights along the entire East Anglian coast. Even when the British got airborne, the Zeppelins could move out of range. Bertie spent long hours patrolling over the North Sea by plane. One night he lost control for a few crucial seconds when his goggles slipped, and crashed into the sea. He was lucky to escape with minor injuries.

Meanwhile, George Senior's sons by his first marriage, Edward and George Junior, were dealing with rapidly changing circumstances at Bournville. All the horses at the factory had been commandeered by the services, and goods could no longer be delivered easily. The trained workforce gradually disappeared. The brothers found it a struggle to recruit people who could operate the specialised machinery, and raw materials were in short supply.

Even under these difficult conditions, the chocolate the company managed to produce proved to be the high-energy comfort food of choice for the armed forces. Far from the anticipated downturn, at the beginning of the war Dairy Milk and Bournville Cocoa were in such demand by the government that the Cadbury cousins had to streamline the chocolate works to gear up production of their two most popular brands. Within two years the seven hundred different chocolate products they were making at the start of the war were reduced to two hundred. They did, however, introduce a much-loved innovation in 1915: a chocolate assortment sold in five-pound cartons and named Milk Tray.

As the naval blockade imposed on Britain by German U-boats and ships intensified, the government ordered the production of essential foods at Bournville. The milk-processing plants at Frampton and Knighton developed for Dairy Milk production were adapted to churn out butter, condensed milk, milk powder and cheese. Sections of the factory at Bournville now made biscuits, dried vegetables and fruit pulp in addition to the core chocolate lines. 'George is having a very anxious time at Bournville,' George Senior confided to a friend on 24 February 1916. 'We have lost some 1,700 out of 3,000 men that were with us when war began . . . to the Army and Navy.' Within a year he reported that another seven hundred of his workers had been moved to munitions factories. It was increasingly hard to find skilled engineers to repair machinery. George Senior remarked that women took over some of the men's jobs, and 'were doing remarkably well'.

The remaining staff contributed to the war effort long after the Bournville bell signalled the end of their working day. Women of the Bournville Nursing Division often worked in local military hospitals at night. The Bournville Girls' Convalescent Home at Bromyard in Herefordshire was transformed into a military convalescent home. Sixty members of the Works Ambulance Class supported the Birmingham division of the St John Ambulance Brigade; their role was to transport the wounded who arrived at Birmingham's Snow Hill station to local hospitals. Bournville volunteers set up stalls at the station and greeted returning soldiers with hot drinks and food. The Works War Relief Committee helped the families of men who were away on active service. The Education Department of Bournville sent books to the troops in the trenches, while the Bournville Dramatic Society kept spirits up with plays on improvised stages.

At a meeting of Fry's directors in Cheltenham, Gloucestershire, in 1916, Edward and George Junior learned that the Frys were seeking greater collaboration between the two firms, especially on pricing. As months of war turned into years, the chocolate companies had been facing shortages of milk and sugar. The government's Sugar Commission introduced quotas in 1916 to ensure that essential foodstuffs had priority. The Frys had decided to concentrate on the cheaper end of the market, but were unable to produce the necessary volume. They were also hit by the loss of exports, which had become prohibitively expensive to insure or simply too dangerous. The struggling house of Fry began to fear being taken over by the predatory Swiss.

It was the Swiss firms that stood to lose the most because of the war. Their home market was small, and production and sales across Continental Europe had been disrupted by the hostilities. The naval blockade made it difficult to secure supplies and to export products to lucrative foreign markets like Britain. By 1916, faced with a shortage of fresh milk, some factories had to be closed. But Nestlé's directors devised a brilliant strategy for getting around the crisis.

Knowing that dairy and chocolate production in many countries was soaring to meet large government orders, the Nestlé directors borrowed on a large scale to set up companies overseas or buy controlling stakes in foreign producers. Through its American branch in Fulton, New York, it acquired interests in firms in North and South America. Its shares in milk-processing companies in Ohio and Philadelphia alone brought it control of twenty-seven factories. Nestlé's production in America was soon five times greater than its entire Swiss production had been before the war. Company directors were in Australia hunting down similar deals, and it was not long before the Swiss set their sights on the lucrative British market. They decided to target Fry.

It was in the English Quaker chocolate firms' interests to help each other keep out foreign rivals. At Cheltenham, Fry, Rowntree and Cadbury agreed to limit production of less essential lines such as confectionery, in favour of basic foods like cocoa and milk. By cooperating, they hoped to pull through the war without falling into Swiss hands.

By early 1917 German submarines were having a devastating effect on supplies of raw materials. Sugar quotas plummeted 50 per cent, forcing the British chocolate houses to adapt the recipes of their remaining core lines to reduce the amount of sugar. Fancy Boxes and other luxury lines disappeared. At Bournville, milk chocolate production stopped altogether. There was a shortage of basic food for the public. 'We are sending 20,000 gallons of milk each week from Frampton to Birmingham as there is a great scarcity in the poorer districts,' George Senior told a friend. Cadbury was forced to abandon its leading line, Dairy Milk.

In Switzerland, thanks to the support of Swiss banks, Nestlé's worldwide acquisitions continued to increase, but so did its borrowing. Loans of twelve million Swiss francs in 1912 rose to fifty-four million by 1917.

◄◖§❀§◗►

Isolated from the barbarity of the European conflict, Milton Hershey continued to thrive. 'Hershey town was so peaceful and serene a community that all the horrors of war one read about seemed strangely out of tune with the course of life,' wrote Joseph Snavely. 'Even the prospect of war with Germany seemed far fetched.'

Nothing seemed to tarnish Hershey's Midas touch. He continued to benefit from his Cuban spending spree as sugar prices rose because of the threat to shipping. By January 1917 German submarines were targeting merchant ships in the Atlantic directly. On 2 April President Woodrow Wilson spoke to a special session of Congress setting out the case for the USA to enter the war. 'German submarine warfare is a warfare against mankind,' he declared. America joined the war four days later.

Almost immediately, the US government placed an order for two million Hershey bars. The firm's sales would soar again, from $10 million in 1915 to $20 million in 1918. Amidst wild rumours that Germany would find a way to sabotage his factory, Hershey set up a Home Guard. 'A finer squad of soldiers never before carried broom handles,' said Snavely, since the workers were not permitted to bear arms.

In contrast to the unrelenting activity at Hershey's chocolate works in America, in Britain, parts of the great Fry's citadel in Bristol fell silent. The firm was struggling. Fearing that Fry – and even Rowntree – might be tempted to sell out to their rich Swiss rivals, Edward Cadbury came up with a radical solution: a merger of the three companies. The large Quaker enterprise that would result, he argued, would 'aim not only to make profits, but also serve the community'.

Continued competition under wartime conditions was wrong, he believed, if together they could give 'better service to the public and create better conditions for our workers'.

On 5 February 1918, at a solemn board meeting in York, the younger members of the Rowntree family, Seebohm and Arnold, argued earnestly in favour of the three Quaker firms joining forces. But Joseph Rowntree was resolutely against it. Perhaps this was a matter of stubborn pride, perhaps it was as a result of his wish for Rowntree's to survive independently in order to pioneer a profit-sharing scheme for its staff. 'The present industrial organisation of the country is unsound,' he said, because it created a division 'between the holders of capital on one side and workers on the other'. He wanted to use the family firm to test out how to 'minimise the evils' of the capitalist system. For Joseph it was a matter of Quaker duty to nurture the guiding light within each member of staff. The real goal for an employer was 'to seek to secure for others . . . *the fullest life* of which an individual is capable'. He believed that the amalgamation of the three firms would result in a money-making behemoth that lost sight of such Quakerly ambitions.

Although Rowntree was against the proposed merger, it joined Fry, Cadbury and other Quaker colleagues for a Conference of Quaker Employers that April. 'War has revolutionised the industrial outlook,' declared the conference chairman, Arnold Rowntree, in his opening address. During this time of crisis there was an urgent need for Quakers to examine ways in which their religious faith could be given 'even fuller expression in business life'. He asked all Friends to consider the Quaker *Book of Discipline*. Citing the section on 'The Stewardship of Wealth', he warned against 'the spirit of greed . . . unchecked by a sense of social responsibility' and urged the conference to find ways to express the Quaker view that 'all human life should be reverenced as capable of the highest distinction'.

Seebohm Rowntree led a session that explored the ethical principles that should be used in determining wages. It was a 'monstrous thing', he declared, that millions of people in Britain still lived below the poverty line, and he encouraged Quaker employers to 'strongly press for State action', arguing that '*every* man should be entitled to a Basic Wage' set at a level that enabled him to 'marry, live in a decent house, and provide the necessaries of physical efficiency for a normal family'.

If a business could not afford to pay such wages, its management should 'very strictly limit' what they paid themselves while they improved the efficiency of their company. George Cadbury Junior led a session on the factors that affect workers' peace of mind: security of employment, quality of environment and so on. Other speakers addressed such topics as industrial injuries, pensions and even the democratisation of industry: schemes for profit-sharing or other forms of partnership, and training to enable a firm's junior staff to aspire to senior positions. The result was a visionary report setting out a template for applying Quaker principles to business life. Years later, after the Second World War, some of these ideas would be enshrined in law with the formation of the Welfare State.

But for all the fine Quakerly ideals, the war brought real issues of business survival. Talks about Fry joining forces with Cadbury continued into 1918. At a directors' meeting in Bournville, George Cadbury Senior opposed a merger of the chocolate houses, but for different and more commercial reasons than his rival in York. 'If we fought off the foreign competition once, I have no fear of doing so again,' he said. His voice may have become frail, but none of his fighting spirit had left him. With ingenuity and drive, he was certain the company could take on its overseas competitors. As part of a merged entity, he felt there would not be 'the same energy thrown by either firm into their business'. Having forged the business from the start, he recognised the significance of that indefinable fighting spirit.

The younger generation of Frys and Cadburys listened to their elders, but remained keen to proceed with a merger. The Great War had exposed fundamental weaknesses in Fry's operation, and Barrow and Edward Cadbury, regarding the matter as 'urgent', feared that the company 'might be tempted to take offers from other quarters'. If Nestlé, for example, bought Fry it would gain an 'insurmountable advantage' in the British market. Over the spring and summer of 1918, the chocolate companies were valued by two different City accountants as a first step towards a potential merger.

By now, letters from Barrow and Edward Cadbury's younger half-brothers were bringing news of a decisive shift in the war. Laurence's ambulance convoy was at the Second Battle of the Marne, seventy-five miles north-east of Paris. At the front line in late June 1918 there was 'great uneasiness', Laurence wrote. 'Everyone knew an attack was coming – but where?' On 15 July, 'the Hun made his greatest and last push over a fifty-mile front'. Later that month, the tables had turned. At a certain point, Laurence reported, the French found there was no one in front of them. They started in pursuit, 'and kept the Hun on the run giving him absolute hell'. It was the first in a series of Allied victories.

Bertie was also beginning to suspect that the war might be approaching its end. Promoted to captain in the new Royal Air Force, he was put in charge of a squadron patrolling for U-boats and Zeppelins. He wrote to his father on 6 August 1918 to report that his 'lucky star has again been in the ascendant . . . Another Zeppelin has gone to destruction.' It all happened 'very quickly and very terribly'. Bertie had been at a concert with his fiancée when he heard that three Zeppelins had been sighted fifty miles north-east of Great Yarmouth. Knowing that there was only one plane available with the necessary speed and climb to pursue the German airships, he 'roared down to the station in an ever-ready Ford, seized a scarf, goggles and helmet . . . and sprinted as hard as ever Nature would let me, and took a running jump into the pilot's seat. I beat my most strenuous competitor by one-fifth of a second.'

As soon as he left the airfield Bertie saw the Zeppelins, and manoeuvred into a position where his gunner, Bob Leckie, could target one of them. The explosive bullets ripped a massive hole in the airship, which 'lighted after a few rounds and went hurtling down to destruction from a great height', wrote Bertie. He and Leckie managed to start a fire in a second Zeppelin, but it was extinguished and the remaining two German airships moved away at high speed.

But now Bertie's luck was out. His engine cut out for a while, and Leckie's gun jammed. 'Our sting had gone . . . I think that half hour, driving through 12,000 feet of cloud in inky blackness on a machine that I had been told could not land at night . . . was the most terrible I have ever experienced.' They later learned that the airship they had brought down was the L70, the best in the German fleet. Among those killed had been Peter Strasser, the head of the German Airship Service. After this raid, there were no further Zeppelin attacks on Britain.

As the guns finally fell silent in November 1918, belief in the supremacy of European civilisation was in question. Thirty countries had battled in a war that involved nations from five continents and left millions dead. Germany and Russia had suffered defeats, and the Austro-Hungarian and Ottoman Empires had ceased to exist. In 1919 the Versailles Peace Treaty brought thirteen million more people and 1.8 million square miles of territory into the British Empire. But the price was terribly high. More than three quarters of a million British fighting men had died. At Bournville, of the 2,000 men who left to join the forces, 218 never came back. In a solemn ceremony, a memorial tablet was raised in their honour.

Those who did come home to Bournville returned to a different commercial landscape. In the final weeks of the Great War, Cadbury and its former great rival Fry had merged in a new holding company called the British Cocoa and Chocolate Company. The Fry family, believing that both firms were of equal value, was shocked when the accountants' valuations were completed and Cadbury's assets were found to be worth three times as much as Fry's. Consequently, Cadbury held the controlling interest in the new company. Shares were issued in relation to the valuations of the two companies. The chairmanship

of the enterprise fell to Barrow Cadbury. Keen to do the right thing by Fry, he valued its shares generously, but the great Bristol chocolate house was now effectively a subsidiary run from Bournville. This was the first of a series of mergers and takeovers that would transform the British chocolate industry over the next century.

It soon became clear that far from being the dynamic partner Cadbury had envisaged, Fry was something of a white elephant. Twenty-six-year-old Bertie Cadbury, honoured with a Distinguished Flying Cross for his exploits as a fighter pilot, had returned from the war only to find there was no room for him at Bournville. Instead he took up an offer from the Fry division of the new company to join it at a salary of £300 a year. He was eager to take on the challenge, but when he turned up at Union Street in Bristol for his first day in May 1919 the state of the firm shocked him.

Everywhere he was greeted by the legacy of Joseph Storrs Fry II's failure to invest in the company. 'Partially manufactured goods were transported by horse-drawn box vans through narrow congested streets,' he wrote home, 'between 24 different factories.' The doors of the many buildings opened directly onto the street, making it impossible to track who came and went, and goods disappeared without trace. Bristol had a pub on every street, meaning that the calm discipline that characterised Bournville was lacking. Equally curious for Bertie, 'It was impossible to shake the then directors out of their complacency and their false sense of security.' The Director of Confectionery and the Director of Chocolate-making were 'hardly on speaking terms'. Since both divisions required large teams of almost identical staff, the costs and inefficiencies multiplied, as did the confusion. Worse still, reported an amazed Bertie, 'Fry's never had much regard for quality, but during the War they abandoned any pretence of maintaining it and they opened up their refiners and let them rip.' The inevitable result was that the company now had a 'terrible reputation' and was producing 'unpleasant' chocolate. These failings were hardly to be wondered at, thought Bertie, 'with a divided and hopelessly inefficient and out of date factory' and 'a sales force split in two'.

As for the Swiss rival that had prompted the merger in the first place, it turned out that, as George Senior had forecast, there was in fact nothing to fear. Nestlé's had continued its bold growth strategy and aggressive spending, acquiring twenty-two more factories in Australia and America in 1920 alone. It had eighty factories worldwide when the post-war downturn kicked in. The combination of plummeting government orders, stunning levels of debt, a crisis in foreign exchange, a rise in the costs of raw materials, and panic selling of Nestlé shares brought the company to a crisis point. In 1921, for the first time in the company's history, Nestlé announced a substantial deficit of one million Swiss francs.

In the years after the war, George Cadbury Senior's health was beginning to fail. As a pacifist he had watched, despairing, as the war consumed its numberless victims, leaving Europe ravaged and the empires that had shaped his world shattered. The evil of war was incomprehensible to him, far removed from his understanding of God. But nothing could destroy his faith. Ultimately he heard only one voice, and focused on the vision that drove him: a liberal vision of toleration, unity and a world at peace.

George's wife Elsie was keen to take an active part in the peace process, and set off to Paris where the Treaty of Versailles was under negotiation. She was particularly interested in the new League of Nations, with its ambitious hopes of safeguarding the future peace of the world. She became a local representative of the League, and worked tirelessly to promote disarmament. When Elsie and George learned there were children starving in Austria after the war, they made arrangements to bring fifteen of them to stay at Bournville. Quaker friends organised 'cocoa rooms' to feed orphaned children in Vienna, and George became known as 'the Chocolate Uncle', on account of the numerous boxes of chocolates he sent out to the children.

Gradually George reduced his duties, and when he appeared in public it was invariably with Elsie at his side. Although still tall and spare, his head held high, George now had a certain fragility, and moved with the careful, measured pace of a man who knows his days too are measured. His eyes 'gleamed with visionary light', as though fixed on the New Jerusalem that had inspired him throughout his life.

On 18 January 1922, George visited the latest developments at the

George Cadbury greets King George V and Queen Mary at Bournville in May 1919.

Selly Oak colleges such as Woodbrooke that they had helped to found, which aimed to advance spiritual and theological understanding between different Churches. He and Elsie were inspecting a new church at Selly Oak when he was taken ill. His doctors advised rest.

Even at eighty-two, George was not a man who would readily succumb to doctors' orders. Confined to the upper floor of the manor house, he could hear the familiar sounds of staff moving around the house, the whirring of the electric lift between floors, the water cascading from the stone fountain outside on the terrace. Peaceful hours passed in his study commenting on the minutes from Bournville board meetings, or seeing visitors with whom he happily engaged in discussing causes about which he was still passionate. His voice was weak and frail, reported a friend, Lady Frances Balfour, who saw him that summer, but his indomitable spirit never left him as he talked enthusiastically of his 'vision of what cities should be'.

From a balcony at the front of the house he could look out over the little empire he had created, still unspoiled. He had always maintained that the view from his window, across the lake and the herbaceous gardens, was a universe in miniature. He loved to watch the children playing in the fields beyond, and that year he and Elsie invited a record number of children for holidays at The Barn and The Beeches. In the other direction, towards Birmingham, beyond the trams of Bristol Road, the rapidly expanding model village of Bournville now had more than a thousand houses and almost 2,000 acres of land, all thriving and at peace.

George Senior was well enough to visit Bournville in October, but any hopes that he would see another Christmas were short-lived. On 20 October he slipped into unconsciousness. There was a brief rally the next day, and he and Elsie seized their chance for one more brief reunion and farewell.

But on 24 October Elsie entered in her diary: 'Go early to see my darling. Breathing difficult all day . . . Just as the Bournville bell signalled the end of the day, one sigh, and he too "went home after work". Such triumph and joy. Much desolation.' She told her friends, 'It was the going home of a conqueror.'

Bright sunlight streamed through the leaves on Bournville village green on 28 October 1922. It was the kind of glorious autumn day that George had always loved, but this was the day of his Memorial Service. Sixteen thousand people had gathered silently to pay their respects. Bouquets of flowers engulfed the Rest House, the memorial on the village green donated by the people of Bournville to commemorate George and Elsie's Silver Wedding, and tributes poured in from around the world, including messages from thousands of people whose lives had been personally touched by George Senior. A young Viennese woman rescued after the war, a criminal whose life was transformed by George's Adult Class, a cripple who had had his first taste of swimming at the farm . . .

The Bournville choir sang the hymns, and the twenty-two bells of the carillon that had once enchanted George and Elsie rang out across the village. At the centre of it all was Elsie, her grief and personal loss staunchly masked behind a veil of formality as wellwishers offered condolences. 'He was very much more than a captain of industry,' declared Dr Henry Hodgkin in the eulogy. 'We think of him . . . as a man of God, whose religion was his life and whose life was his religion . . . It seemed as if none lay outside his great heart. The world was his parish. He had a universal spirit.'

People mourned not just the loss of a pioneer, but of all that George

Cadbury had been: the practical mystic whose visions of utopia, of garden cities and factories, whose love of nature and the simple things in life, whose overpowering faith and desire to do good had informed every aspect of the enterprise he created. Above all, said Hodgkin, 'he was a man who loved. The inspiration of his life was love . . . Can any epitaph be more worthy than this – "He loved greatly"?'

Perhaps George Cadbury would have marvelled at the fuss. He believed he was merely doing his job, and was fortunate to be in the position to act on his ideals. His faith had always pointed the way. He was unable to give up hope, even when faced with the meaningless horror of the Great War. His conviction that a person's soul 'lived or perished according to its use of the gift of life' had guided him from the outset, and never wavered. Neither he nor his brother Richard, innocents at business when they took over the factory from their father, had any conception of the great riches that following this course would provide. But business for them had always been just a means to an end. As Quakers, they knew the rewards were for their fellow men and for the glory of their God.

A similar scene would be played out in York in February 1925, when George Senior's friend and rival Joseph Rowntree died at the age of eighty-eight. Right to the end, according to his private papers, he was puzzling over the question of poverty, and whether, through rigorous scientific enquiry, it could be made a thing of the past. In York, as in Bournville, people mourned not just the passing of the man, but of everything he symbolised that had brought such unexpected good to the world of business, on such a large scale.

Although the two Quaker pioneers had died, they left behind enterprises in York and Bournville that were larger than themselves. People spoke of the 'spirit' of the Quaker chocolate firms, as though a mantle had gently enveloped each one, granting it a life and character of its own. Each seemed charged still with the mysterious purpose of the men who created it, as though even the bricks and mortar were transformed, bent towards some stoic Quaker quest. Would such dedicated altruism falter now that its leading lights were gone? The spiritual values that had united and guided all aspects of George Cadbury's and Joseph Rowntree's lives had not seemed out of place in the Victorian and Edwardian eras. But what hope was there that their creations would live on in the changed landscape of the post-war world?

PART FOUR

16

I Pray for Snickers

The Roaring Twenties arrived, exorcising some of the horrors of the Great War. The growing wealth of the United States meant that many households now owned a car, a radio, sometimes a fridge or even a toaster. This new world of consumerism and easy credit was far removed from the frugality and self-restraint of the Quakers. Only a few years had passed since the death of George Cadbury Senior, but the chocolate companies were about to be thrust onto a different planet: Planet Mars.

Forrest Mars was born in the USA in 1904, but after the separation of his parents he was sent to live with his grandparents in the remote mining community of North Battleford, Saskatchewan, Canada. His entry into the world of confectionery happened by chance in 1923, when he was reunited with his estranged father, Frank. Frank was running a sweet company called Mar O Bar, which was not well known beyond its locality in Minnesota. Forrest urged his father to think big. Surely they could come up with a new type of confection that they could sell from coast to coast, like Hershey?

Forrest had undoubtedly been aware from his childhood of the potential of a particular type of confectionery made by combining cheaper ingredients with chocolate. In Canada, handmade bars of maple fudge or marshmallow coated in chocolate had been on sale for many years. This type of confection was becoming known in the trade as a 'countline', since chocolate-coated bars were usually sold by 'count', not by weight, like a solid block of chocolate. It occurred to Forrest that while leading manufacturers such as Hershey and Cadbury were focused on producing solid chocolate bars, no one was mass-producing countlines.

Soon after his reunion with his son, Frank Mars developed a countline of his own: a bar of nougat and a layer of caramel wrapped in a thick coating of chocolate. The bar was satisfyingly chunky compared to a Hershey chocolate bar, and could be sold more cheaply since the confectionery ingredients cost less than cocoa. Equally important, the chocolate coating kept the nougat filling fresh, making it possible to ship around the country. Frank Mars successfully approached Hershey to supply chocolate for the coating. By early 1924 he was ready to launch his new bar. He called it the Milky Way.

The Milky Way was an overnight sensation. In one year, Frank's dreams came true. Sales at the Mar O Bar Company leapt from $72,800 to $792,900. Frank was ready to expand beyond Minnesota, and built a stylish new factory in Chicago. Forrest joined him, and nothing could stop them, not even the financial drought of the Great Depression. In 1930, a year when 26,000 American businesses failed, and 10 per cent of the nation's population was out of work, Frank and Forrest Mars rode out the crash on a wave of innovation. That year they launched a peanut, caramel and chocolate sensation called Snickers, reputedly named after the family horse. The 3 Musketeers bar soon followed. Just eight years after father and son had reunited, Mars, as their company was now known, was the second-largest confectioner in America.

It was a family rift that launched young Forrest on his own path to success. In 1933 father and son had a serious quarrel, from which their relationship never recovered. According to an article in *Fortune* in 1967, Frank sent Forrest packing with the words, 'This company isn't big enough for both of us. Go to some other country and start your own business.' Forrest walked out of the Chicago factory with the foreign rights to the Milky Way, and never saw his father again.

After a brief spell learning the secrets of Swiss chocolate, he came to England. His plan was to enlist the help of the benign and unsuspecting Quaker gentlemen at Bournville to launch a chocolate company of his own. Unencumbered with religious values, he was free to bring his own driven business style to what he regarded as the quaintly philanthropic and paternalistic world of English chocolate-making.

Bournville works outing, 1920.

The steps that would transform the private Quaker firm of Cadbury into the largest confectionery company in the world began in the 1920s, as successive generations of the family grappled with a series of challenges from foreign rivals. After the First World War, it was the formidable Swiss adversary Nestlé which appeared to pose the greatest threat to the British chocolate firms. At Bournville, reinforcements were needed to help with the expected hostilities. A younger generation joined the family team. Barrow Cadbury, chairman of the British Chocolate and Cocoa Company, brought in his oldest son, Paul, while Edward and George Junior had the assistance of their younger brothers Laurence, and Bertie – who joined the Fry division.

In the early 1920s much of the chocolate manufactured at Bournville was still handmade, but the Cadbury cousins recognised that the future lay with the mass manufacture of fewer lines. Supply rationing during the Great War had forced efficiencies on Bournville. The cousins knew it was possible to produce volume more cheaply, which in turn would enable them to sell their products more cheaply. This seemed an imperative for survival. If foreign rivals succeeded in gaining an edge in mass production, the British Chocolate and Cocoa Company could face a slow slide into oblivion.

The cousins embarked on the largest transformation of both Fry's and Cadbury's chocolate works yet. In Bristol, Bertie recommended that Fry abandon the centuries-old Union Street site beloved of Joseph Storrs Fry II and invest in a modern factory outside the city. In 1920 he and the Fry directors walked along the railway lines out of Bristol until they found the ideal location, near Keynsham. They bought over two hundred acres bounded by the River Avon and the main Bristol to London railway, and embarked on a massive building programme to create their new factory, named Somerdale.

At Bournville, Laurence Cadbury, as a skilled engineer, supervised the designs of automatic production lines that were tailor-made for virtually every stage of chocolate-making. The old Victorian blocks were knocked down to make way for the chocolate factory of the future, which eventually covered eighty acres. Over the course of the 1920s, Laurence watched with satisfaction as the new five- or six-storey buildings with their imposing red-brick façades replaced their darker, smaller Victorian predecessors. Laurence married Joyce Matthews in 1924, and soon a young family was on the way. Julian was born in 1926, followed by Adrian in 1929.

Nestlé was slower to recover from the effects of the Great War than anticipated. Its share price crashed in the early 1920s, when it was revealed that the company had debts approaching 293 million Swiss francs. In York, the Rowntrees failed to find dazzling new

The dining block at Bournville, 1927.

confections to win over the public. But the Cadburys had a string of successful innovations, including the extra-light Flake, the Fruit and Nut chocolate bar, the honeycombed Crunchie (made by the Fry division) and Crème Eggs for Easter, filled with fondant, marshmallow or marzipan. The Bournville chocolate-makers were soon to be challenged, however, by an unexpected newcomer to the chocolate wars, who would change the confectionery business for all time.

In 1933, Cadbury's sales team became aware of a fledgling company operating from a tiny flat in Slough, twenty miles west of London. It was run by an American, Forrest Mars, who approached Cadbury direct for a supply of chocolate to coat his new confectionery. To his delight, Cadbury's coatings department agreed. Even today Laurence's second son, the former company chairman Sir Adrian Cadbury, smiles and shakes his head as he looks back on that decision taken by his uncles. 'Why ever did we do that?' he says.

In England in the summer heat of 1933, Forrest Mars set out to reconfigure the Milky Way into a chunkier, more satisfying bar. He hired four assistants, and they set to work in an old building in Slough. On one occasion he was nearly ruined when rain leaked through a roof onto sacks of raw materials, but Noah's Flood would not have deterred Forrest Mars. At long last he felt he had developed the perfect formula, combining a layer of soft caramel and a wrapping of the ever-popular Cadbury milk chocolate around a whipped fondant centre. The resulting Mars Bar proved to be one of the most successful snacks of all time. In its first year it sold two million bars.

In York, a new director at Rowntree, George Harris, was paying close attention to the novel idea of mass-producing countlines. The company had struggled during the Depression, and for a time the chairman, Seebohm Rowntree, had feared for the survival of the business. Harris recognised that Rowntree could never challenge Cadbury's supremacy in block chocolate: Dairy Milk was so dominant it was outselling Rowntree's milk chocolate ten to one. But could Mars's countlines offer a way forward?

Under Harris's pioneering lead, the research department at Rowntree was kept very busy. Hundreds of tests were carried out on a new chocolate box to compete with Cadbury's leading King George assortment. Finally they came up with Rowntree's Black Magic, which launched in 1933 to immediate success. Forrest Mars introduced the

Milky Way bar in Britain in 1935, followed by Maltesers in 1936, but Rowntree's also had another novelty: a little bar called Chocolate Crisp, launched without advertising in September 1935. It was only a wafer covered with chocolate, a wisp of a thing, but it proved to be an instant winner, and it soon got a catchy new name: Kit Kat. That same year Rowntree surprised the market again with the bubbly Aero, which teased the palate with its deceptive lightness. In 1936 came Rowntree's Dairy Box, full of soft-centred chocolate temptation. Cadbury countered with intense promotion of its popular Milk Tray box, and launched a new assortment known as Cadbury's Roses – supposedly named for George Senior's passion for that flower. But George Harris had yet another trick up his sleeve. Soft chocolate beans coated in crispy shells of coloured chocolate were repackaged in an eye-catching cardboard tube and sold as Smarties in 1937. To create so many distinctive brands in such rapid succession was almost unheard of. At last Rowntree's fortunes were turning as the chocolate wars intensified.

Into the fray came the Swiss giant Nestlé, making the competitive skirmishes of England's Quaker gentlemen look like a children's game. Nestlé's collaboration with the Swiss chocolatiers Peter-Cailler-Kohler had hit its stride in the 1920s, and made chocolate its second most important product. In 1929 the two companies had sealed their alliance with a full merger that brought six more chocolate factories in Europe alone under Nestlé's control. As a result, in that year, when stock markets crashed around the world, Nestlé's profits actually rose from twenty-three to thirty million Swiss francs. During the 1930s Nestlé's block chocolate, with the aid of an impressive armoury of promotional tools – prize schemes, coupons, collecting cards for children – was seizing new ground in British grocers' shops.

But the Cadbury cousins were ready for the new challengers. By 1935 the modernisation of Bournville was complete. The new factory was unlike anything that had been seen before: a vast mechanical palace where sugar and cocoa were metamorphosed into 'food for the gods'. Every process, from moulding to wrapping, was fully automatic. The Cocoa Block alone had over a mile of conveyer belts that carried the beans from trains at one end and churned out cocoa tins ready for loading at the other. The scope of the reorganisation was so huge that only a quarter of the works that had existed in the days of George

Senior remained. In the late 1930s, a million Dairy Milk bars and two million Chocolate Assortments rolled off Bournville's lines every day. The sheer scale of the operation meant that the Cadbury cousins could afford to cut the price of Dairy Milk by a staggering 70 per cent. The increased automation did not lead to redundancies, as George Senior had feared: staff numbers rose by 2,500 in the interwar years, reaching 10,000 by 1938 at Bournville alone, and a Quakerly concern for the well-being of the workforce continued.

Women's gymnastics at Bournville, 1937.

In the late 1930s, Rowntree was promoting its Aero bar in a way that challenged Dairy Milk head on, claiming it was more easily digested than 'old fashioned milk chocolate'. It also took the unusual step of trying to patent the bar's production method, although some of the processes involved were already in use in the industry. Cadbury joined forces with its rival Nestlé, which also objected to Rowntree's tactics, to mount a legal challenge.

Quaker solidarity and allegiances shared over a century appeared to have evaporated in the face of the intense competition to win market share. Seebohm Rowntree, perhaps uncomfortable with the strained situation, chose this time to retire from the Rowntree board. The possibility that the dispute could be settled by the Society of Friends was considered. This was, after all, how Quaker firms had resolved their differences in the past. But such a meeting never took place. The Society had ceased to be a relevant forum for business negotiation. Instead, all parties resorted to lawyers, and it was late summer 1937 before an agreement was reached that required Rowntree to back down on both its patents and its promotional claims.

Faced with the cut-throat conditions of an increasingly saturated market, the moderating Quaker influence declined. In York, the Quaker voice diminished in the Rowntree's boardroom as independent managers replaced family directors who left or retired. According to the business historian Robert Fitzgerald, Rowntree family members, with their 'aesthetic form of entrepreneurship', no longer had preference for seats on the board. At Bournville, the Cadbury cousins found themselves straddling a widening gulf between business pressures and religious values.

George Senior had pointed to a way forward when he created the Bournville Village Trust, and many of his and his brother Richard's children built on this legacy. As Quakers, they did not wish to accumulate large personal fortunes; instead they created trusts or made gifts on a formidable scale. Barrow Cadbury, Richard's oldest son, gave away much of his inheritance, and fulfilled his father's ambition of creating an institute for the local Adult Schools in Moseley Road in Birmingham. He and his wife Geraldine were local magistrates, and set up a Children's Court and Remand Home to help young offenders. In the 1920s they used their shares in Bournville to establish two large charitable trusts which they particularly wanted to help missionaries and those who had spent their lives helping others. Barrow and Geraldine also gave several former Cadbury houses to the community, such as the home for sick children at Cropwood and the Copeley Hill Hostel for young adults.

Barrow's brother William, and his cousins Edward and George Junior, continued this tradition by setting up sizeable benevolent trusts. Edward and George Junior also donated substantial chunks of land: two hundred acres on the Lickey Hills near Bournville went to the City of Birmingham, and the four hundred acres of the Chadwich Manor Estate, near Bromsgrove in Worcestershire, went to the National Trust. They also gave buildings and funds to Birmingham University and the Queen Elizabeth Hospital Centre at Edgbaston.

For Richard's idealistic youngest daughter Beatrice, born in 1884, the desire to give up material possessions went to extremes. She had inherited a considerable shareholding in Bournville, yet as a missionary saw it as her duty to 'mitigate the acute suffering caused by the capitalist order of society'. She and her Dutch husband, Cornelis Boeke, felt they should give away everything, and tried to live with no money at all. When her brother Barrow visited her and her eight children, he was shocked to find them living in tents. He insisted on setting up a trust to provide the young family with a modest home in Bilthoven in Holland.

The 'capitalist order' that Beatrice questioned was bringing a stunning array of luxuries to the consumer. In Britain by the late 1930s, chocolate was no longer a rare indulgence but a routine purchase. Almost everyone was eating chocolate, compared to just 3 per cent of the population in 1900. Cadbury's Dairy Milk, at '2 oz for 2d', as the advertising slogan had it, was the most popular chocolate bar by a large margin. Cadbury now had factories in Tasmania, Canada and New Zealand. Bournville, at the hub of it all, the largest and most modern chocolate works in the world, was nominated by one British newspaper as one of the Seven Wonders of Britain.

The Cadbury cousins were not alone in their global vision. Forrest Mars, fast rising to become Britain's third largest confectioner, had ambitious dreams of furthering his chocolate business with a scheme that would challenge the huge Hershey empire. It took the threat of the Second World War to spur him into action. Just before the war started, the British government implemented plans to tax all resident foreigners. Forrest Mars chose this moment to leave England.

❦

With the outbreak of war, Bournville and Somerdale chocolate works, gleaming showcases of mass manufacture, were examined as potential sites to make munitions. Somerdale, gutted of its state-of-the-art chocolate technology, was hastily converted for the Bristol Aeroplane Company. At Bournville, firms that supplied military needs such as Lucas and Austin moved in. Gone were the lines of rosy girls in virginal white tending to the nation's sweet tooth, replaced by the gunmetal colours of war.

'It was a weird place,' recalls Adrian Cadbury, ten years old when the war started. 'It was covered in green netting and there were workers on the roof scouting for enemy aircraft . . . Inside it was strange because parts of it were exactly as they had been and others transformed.' The Chocolate Moulding Department made gun doors for Spitfires and cases for aeroplane flares. The Chocolate Packing

Department created gas masks. The Metals Department produced aircraft parts. Elsewhere at Bournville millions of jerry cans, petrol tanks for Spitfires and Lancasters and gun mounts were in production. Every resource possible was diverted to the war effort. Bournville's games fields were dug up to grow vegetables as part of the 'Dig for Victory' campaign. Sheep grazed on the village green. Older workers at Bournville formed their own Home Guard.

As a target for German air raids, Bournville was camouflaged with netting.

To his immense frustration, Adrian was not permitted to work on the production line but had to stay at home with his young sisters, Veronica and Anthea, and the new arrival, a baby brother named Dominic. His uncle Edward was running a much-reduced chocolate works. Initially there were large government orders to meet for the army, navy and air force. But as milk supplies and the movement of foodstuffs came under government control, and imports of cocoa and sugar were threatened, core lines could no longer be made. To eke out precious sugar supplies, Economy Red Label Drinking Chocolate with added saccharin was introduced in July 1940, in emergency grey wrapping. In place of Dairy Milk came Ration Chocolate in 1941, a dull and dreary substitute made with dried skimmed milk powder.

First-aid class at Bournville.

A Cadbury 'cocoa caravan' brings comfort to a bombed-out city street.

With the confectionery ration gradually reduced to two ounces per person per week, the chocolate works struggled.

At thirteen, Adrian was finally permitted to take on a 'proper job' at Bournville: 'My first job in the morning was as a post boy and my last job at night was to do the blackout in the chemists' department. The blinds were huge, and there were curtains to put in place before we left for the day.' As the military operation expanded during 1941, Adrian remembers huts being assembled by the canal, where workers packed explosives into anti-aircraft rockets.

On 7 December 1941 Japanese bombers sank or damaged eighteen American warships at Pearl Harbor in Hawaii. The next day America joined the war – an event Milton Hershey had long anticipated. As young men left the chocolate works at Hershey to enlist, the president of the company, William Murrie, stepped up production of specially designed chocolate bars stuffed with vitamins known as Ration D and Ration K. With one billion bars supplied to US troops over the next four years, it is hardly surprising that Hershey's profits nearly doubled during the war, reaching $80 million in 1944.

Milton Hershey was once again at the heart of a chocolate bonanza, but his time was running out. His heart was weak, and increasingly he was confined to two rooms he had kept for himself at High Point, surrounded by photographs of Kitty. His nurses found him quixotic and playful. Sometimes recalling a touch of dash from his Cuban days, he might dress up smartly, put on sunglasses and a hat and have a flutter at the racetrack or in the casinos. With the carelessness of someone who no longer worried about the outcome, he ignored his doctors' instructions and enjoyed rich food and champagne whenever he was in the mood.

He still delighted in visiting the orphan boys of the Hershey Industrial School, and had secretly transferred his stock of $60 million in his chocolate company into a trust to benefit them. 'Too much money is an evil influence,' he told the *New York Times*, which broke the story on 18 November 1923. 'Money spoils more men than it makes.' The Hershey Industrial School had expanded rapidly, and now educated over a thousand boys who lived in surrounding farms and homes. Many had been rescued from extremes of poverty and hardship, and felt deep gratitude to Milton Hershey. On one occasion when he rose to say a few words to the school assembly, the room erupted in clapping and cheering from over seven hundred boys. Hershey was overcome, and tears rolled down his cheeks. A man of few words at the best of times, he could not speak at all, but simply waved his exit.

When Milton Hershey died in October 1945, at the age of eighty-eight, 10,000 people passed by his open casket, mourning the loss of a 'fairy godfather'. Inspired by the Quakers and other philanthropists he had given away most of his wealth, and had earned legendary status in America. Eight graduates of the Hershey Industrial School served as pallbearers in a procession that stretched almost a mile to the Hershey cemetery. Everyone in town wanted to pay their respects as his casket was taken to the vault where he was reunited with his beloved wife. 'When we thank God for his life,' said the Reverend John Treder in the eulogy, 'it will be for the vision it held of a better life for other men.' For most there was a certainty that his utopian dream would survive, and his spirit live on.

The loss of such a giant among the heroes of the industrial age in America left a vacuum, and there were plenty whose sights were set on filling it. First among them was Forrest Mars. Curiously, when he returned to America he did not use his father's company as his springboard. Since his father's death in 1934, his stepmother Ethel and his half-sister Patricia had been running Mars in Chicago. In arguably his cheekiest move yet, in August 1939 Forrest Mars went to Hershey and asked to see William Murrie. He proposed that Murrie's son Bruce should join him in a new chocolate venture. As for the product, he had a plan: chocolate drops coated in a brightly-coloured sugary shell. They decided to call the new sweets M&Ms – for 'Murrie and Mars'. At his factory in Newark, New Jersey, Mars was able to benefit from being under the nurturing wing of his rival, Hershey. William Murrie ensured that his son was never short of cocoa and sugar, and helped the new company to thrive.

In 1945 Forrest's stepmother Ethel died, leaving him half her shares in his father's chocolate company. At last he had a stake in the Chicago firm – roughly one third of the business. If he could unite his own American business with the firm his father had founded, he would create a giant Mars empire. And he had all the momentum of the US market behind him.

Nestlé too had grasped the shift in the balance of power from Europe to the USA. Having learned lessons from the Great War, in the Second World War Nestlé's directors divided the company's management team in two. A core team remained in Vevey to run the European division, while the chairman and his staff moved to Stamford, Connecticut, to expand Nestlé in North and South America. They flourished, thanks to the remarkable response to a very novel product: Nescafé. Instant coffee proved an immediate success in the land of fast food, and Nestlé emerged from the war with its enterprise firmly established in both Europe and America.

At Bournville, recovery from the war years was slow. Everything was rationed, from food to furniture. Laurence Cadbury, who took over as chairman from his half-brother Edward in the final months of the war, managed the austerity years at home by growing the Cadbury businesses overseas. Their works in Canada, South Africa, Australia, New Zealand and Ireland were expanded in the eight years before rationing in Britain ended at last in 1953.

Laurence took advantage of the new opportunities at home. A new factory at Moreton, in the Wirral, was soon making three million chocolate biscuits a day, while Somerdale was expanded to deal with demand for a new bar known as Picnic. At Bournville, Cadbury's Buttons rolled off the production line in 1960, soon followed by redesigned chocolate bars and Roses, and a new biscuit confectionery called Bar Six. Technology for the production of Easter eggs was improved, and night shifts were introduced.

Hard on the heels of the new wave of consumerism came a new medium to promote it: television. Commercial television was launched in Britain in September 1955, and it was to have a profound effect on the confectionery industry. Cadbury was on air the very first evening, with an advertisement for drinking chocolate. Rowntree was equally quick off the mark, with a succession of memorable campaigns: 'Have a break – have a Kit Kat', 'Don't forget the Fruit Gums, Mum', and 'Polo – the mint with the hole'.

The slick one-liners of TV advertisements favoured brands with a simple, playful message – perfect for Mars and Rowntree's large portfolio of countlines. Within three years television accounted for more than 60 per cent of all chocolate advertising. As countlines began to take a share of the market for block chocolate, Paul Cadbury and the marketing team fought back with an innovative portfolio of advertisements. Models were chosen to promote the superlight Flake, while a James-Bond-style action hero overcame a succession of hazardous obstacles to deliver a box of Milk Tray, with the slogan 'All because the lady loves Milk Tray'.

In 1959 Paul Cadbury took over as chairman from his cousin Laurence. Paul had led some of Cadbury's most famous sales campaigns, including '2 oz for 2d' and the enduring 'Glass-and-a-half' slogan, both for Dairy Milk. But Forrest Mars's next move took even him by surprise.

Mars had excelled at creating countlines in which the chocolate was mixed with other ingredients, but in 1960 he used television for a direct assault on Cadbury's block chocolate lead by relaunching a block chocolate bar of his own: Galaxy. It was ushered in with the largest British television campaign for chocolate yet. Cadbury hurried to reposition Dairy Milk by changing the size of the bar, accompanied by heavy promotion, but it was clear that television was changing the rules. One good TV campaign could undermine decades of customer loyalty in a matter of weeks.

There was one way to fight back. In a reversal of the views of nineteenth-century Quaker founders such as Joseph Fry and Joseph Rowntree, by 1960 the chocolate and confectionery firms were among Britain's biggest advertising spenders. Cadbury, at £3.2 million a year, was fifth; Mars, at £2.9 million, was sixth; Rowntree, at £2.8 million, was seventh; and Nestlé, at £2.3 million, was eighth.

In 1961 Paul faced another critical issue much closer to home. There was growing pressure to take the company public. Cadbury had been a private Quaker concern for 140 years, and such a move would be one of the most significant transitions in the firm's history. Adrian Cadbury soon learned of this predicament. After completing an

economics degree at Cambridge he had worked his way through the
ranks at Bournville, and had recently joined the board as a director.

'The problem,' he remembers, 'was that we had family shareholders,
particularly the Fry family, whose capital had been tied up in the
business since 1919, and they did not have a market for their shares.'
By the 1960s this was becoming a significant concern. Although there
were about ten Cadbury family members directly involved in the
business, there were no Frys; and in the wider Fry and Cadbury family,
several hundred people held shares in the firm. There was a desire
to 'do the right thing by Fry's', says Adrian, and by the growing number
of Cadbury family members who could not get access to their capital.

'There is no way the handful of family members who were in the
business could have bought out those who were not,' says Adrian's
younger brother Dominic, who was studying for an MBA at Stanford
Business School at the time. More than 50 per cent of Cadbury shares,
he explains, were owned by benevolent trusts such as the Barrow
Cadbury Trust and the William Cadbury Trust. Because Barrow and
William Cadbury had given away most of their wealth, their sons,
Paul and Charles, 'were never what you would call very rich people,
and could not possibly have bought out the hundreds of Cadbury and
Fry members who did own shares'.

The family members who did want to sell were keen to have an
objective valuation of the cash value of their shares. In the case of a
private company, an independent firm of accountants typically sets
the price per share. 'But this was clearly different from having an
open market in which they knew that the price they got for their
shares was the way the market valued it,' says Adrian. To solve this
problem, and to allow Fry and Cadbury family members access to
their capital, Paul and the Cadbury board agreed to go public. The
British Cocoa and Chocolate Company was floated in 1962. For the
first time since its founding in 1824, the enterprise was no longer
under direct Quaker family control. The management now had to
report to independent shareholders, who demanded a broad-based
and profitable business.

'There were teething problems,' says Adrian. 'For a time the
Cadbury board continued running as if we weren't a public company.'
For years, Paul had sidestepped the top management in the British
Cocoa and Chocolate Company board, and instead used the subsidiary

Cadbury board to run the business, but that situation could not continue after the flotation. Adrian knew the company needed to have a proper Annual General Meeting, and formal procedures for reporting to the board. 'I was concerned,' he says. 'We simply had to recognise that things had to change.'

Just how much they had to change is shown by what was happening in a rival's boardroom 2,000 miles away, on another continent. Forrest Mars had battled for complete control of his father's factory, despite opposition from his stepmother's family. Tragically for the family, the turning point came when his half-sister Patricia was diagnosed with terminal cancer. At last she agreed to sell him her shares, giving him two-thirds of the business, but he wanted it all. This was his father's life's work, and strangers had a voice on the board. Forrest worked unceasingly to persuade other shareowners to sell to him, and he finally achieved full control in 1964, at the age of sixty. It was the end of two decades of struggle.

What happened next was reported by Harold Meyers in *Fortune* in 1967. Forrest, whose obsession with the family business was now 'bordering on religious zealotry', called a meeting of leading executives shortly after he won control of his father's Chicago factory. Forrest Mars did not walk into the boardroom, 'he charged in', wrote Meyers. The senior staff in the room were a little unsure what to make of him. His appearance made no deference to current fashions: 'his English suit had wide lapels and his tie was unstylishly wider still'. Yet he communicated his intensity with a power that was unsettling. After a brief and wary exchange, he presented his vision for Mars confectionery. 'I'm a religious man,' he declared. He dropped from his chair to his knees as though he were in a church pew. The staff watched mesmerised.

'I pray for Milky Way,' he said.

Long pause.

'I pray for Snickers . . . '

No one said a word.

The message was clear: his prayers were for profit. He expected nothing less than the same religious fervour from his staff. It was a very different scene from those of almost a hundred years earlier, when George Cadbury Senior asked his Bournville staff to join him in prayer to seek guidance on a difficult business issue. But Forrest Mars confirmed what everyone already knew: money was the new religion.

17

The Quaker Voice
Could Still be Heard

It would fall to a fourth generation of Cadbury brothers to manage a period of spectacular change. In 1965, three years after taking the company public, Paul stepped down and handed the reins to Adrian, who at thirty-six became the firm's youngest chairman. His younger brother Dominic had just started working at Bournville, and it wasn't long before the two brothers had first-hand experience of Forrest Mars's tactics.

'I met Forrest Mars Senior in London and he offered me a job right in front of Adrian,' said Dominic, smiling as he recollected the scene. Mars and his wife Audrey, who were friends of Dominic and Adrian's parents, were holding a drinks party at the Dorchester, one of London's smartest hotels. 'They had the best suite of rooms, at the top. We were all invited because as families we were close.'

A slight hush followed Forrest's job offer. Adrian, forever the diplomat, was quick to step in and defuse the situation.

'Whatever you offer Dominic, Forrest, we will pay him more,' he said mischievously.

Dominic knew that was not true, but he had no intention of working for his brother's most dynamic rival. 'Forrest Senior was that sort of chap,' he says. 'He was an interesting man. He had a real wicked smile about him. We knew his employees saw a different side of him. He could be a complete tyrant.'

While Dominic started work in South Africa, Adrian was grappling with turning a 140-year-old private Quaker business into a modern public company. Forrest Mars, who claimed to be 'a practitioner of scientific management', pioneered the idea of the open plan office,

Adrian and Dominic Cadbury, grandsons of George Senior, were the fourth generation of brothers to lead the firm.

and took staff out of their comfort zones. By contrast, in the Quaker firm there was a strong collaborative style of management that had evolved over more than a century. Adrian found this was 'wonderful at involving everyone', but could make decision-taking slow.

He knew the organisation had to develop clear lines of personal responsibility. Above all, he wanted to focus on strategy: reducing the

The Bournville works councils dealt with everything from timekeeping bonuses to discipline.

dependence on cocoa, improving the geographical spread of the business, and expanding the foods division. He soon found an opportunity.

The soft-drinks company Schweppes had come a long way from its origins in Geneva in 1783, when the watchmaker and amateur scientist Jean Jacob Schweppe refined a method to carbonate water. By the 1960s the company had grown into an international concern specialising in fizzy drinks such as bitter lemon, ginger ale and tonic water. The management of Schweppes was particularly keen on a merger, as they feared a takeover, and in 1968 Lord Watkinson, the chairman, approached Adrian.

'Even then, size was the only real protection against being taken over,' says Adrian. He was less concerned about a takeover than Schweppes, in part because of the Cadbury benevolent trusts, which in the 1960s held such a large share of the company that potential buyers were put off. And not without reason. The sons of those who had set up the trusts also worked in the business. 'There was a seamless link between the people running the trusts and the business that kept Quaker values at the heart of the company,' Adrian explains. 'This was wrongly perceived as a barrier to takeover.' Adrian knew full well that the trustees' legal duty was to the trusts, not to the company, and trustees could not have protected Cadbury against a takeover bid if it was to the advantage of the trusts.

For the traditionally Quaker firm of Cadbury, there was just one obstacle to a merger with Schweppes: its soft drinks were used as mixers in alcoholic drinks. Schweppes even distributed an alcoholic brand, the liqueur Dubonnet. Was this compatible with the aspirations of Cadbury's founders? Several older members of the family voiced opposition. Adrian's great-grandfather John Cadbury and his wife Candia had been passionate advocates of the temperance cause; indeed the very reason for the creation of the business had been to develop cocoa as a nutritious alternative to alcohol.

Adrian spent a lot of time talking to those in the family who had concerns. He explained that together the two companies would have greater resources, and their geographical spread complemented each other. Both companies had Foods Divisions, which if combined would become more of a force in the grocery trade. This was about 'extending our reach in a fast-moving world'. He pressed ahead with the merger, and in 1969 became joint managing director and deputy chairman of Cadbury Schweppes, under Lord Watkinson as chairman. Cadbury Schweppes was a global giant, with turnover approaching £250 million, a third that of Nestlé.

As Cadbury Schweppes competed for overseas markets, there was one crucial difference between it and its rival Nestlé that was easily overlooked in the small print of the Swiss company's governance rules. Nestlé had continued to strengthen its international presence through a policy of acquisition, and was emerging as one of the world's leading firms. 'To protect these characteristics,' wrote Jean Heer in the company history, 'the management proposed the issuing of registered shares, whose registration would be subject to authorisation by the board of directors and be limited to Swiss citizens.' This, it was argued, would ensure 'stability and balance' in the way the company shares were distributed 'both inside and outside Switzerland'. Unlike its British chocolate counterparts, Cadbury and Rowntree, Nestlé effectively had a two-tier share system, giving it protection against foreign takeover bids.

At Bournville, the results of the merger with Schweppes were evident as a succession of new temptations rolled off the production lines. The chocolate-and-caramel Curly Wurly was followed by other innovative countlines such as the Double Decker and Caramel.

A taste panel at Cadbury in the 1960s.

Dominic Cadbury, working in South Africa, saw the effects of the merger on the overseas divisions. At the time, he says, Cadbury was operating as a British firm with businesses 'dotted around the world', 'rather than a proper international company'. After the merger, the management aimed to create a unified overseas policy: 'There was pressure to show shareholders that there were real benefits of the merger.'

Enforcing a unified global policy, however, was not straightforward. The idiosyncrasies of the Indian market highlight the difficulties. India was a potentially vast market, but Cadbury had to adapt if it was to maintain a foothold there. At first there was no Indian dairy industry. Adrian remembers visiting the new dairy herd created by Cadbury Schweppes on a farm near Poona, south of Bombay. The Indians running the farm were determined to keep hens as well as cows – despite eyebrows raised by Cadbury management. The hens soon brought rats, and then came cobras to keep the rats down. 'The Indians didn't worry too much about them,' Adrian recalls. But they did worry about the cows when they grew too old to produce milk. It was not acceptable for them to go to slaughter. The image of cobras sleeping off a meal of rats among a fast-growing herd of non-productive sacred cows was not ideal for a streamlined multinational firm. Small wonder that the finance director could see no prospect of a viable Indian business.

The suffocating heat of the subcontinent proved to be another obstacle, and much juggling and experimenting with recipes took place before chocolate that did not melt like wax could be made. The production team eventually developed new lines that were less affected by heat, such as toffee-covered chocolates called Éclairs.

Just as Cadbury Schweppes was making headway in India, the government introduced restrictions against foreign companies. 'We had a period where we could not take dividends,' recalls Adrian, 'and a period where we had to sell 51 per cent of the company to Indian shareholders. We changed our name to Hindustan Cocoa Products.' Under this name the company found itself in the surprising position of being protected by the Indian government, which decided that no more foreign brands were allowed into the country. Coca-Cola and Pepsi could not get in, nor could chocolate rivals such as Mars and Suchard. Hindustan Cocoa Products, however, was regarded as 'effectively Indian'.

Taking the long view would pay off, as Cadbury later regained control of the company, which was renamed Cadbury India. In time more shops in India were air-conditioned, and the sale of block chocolate began to rise. Temperature-tolerant countlines were established before rivals like Mars arrived, and Cadbury India was soon exporting to other hot climates like the Middle East, Singapore, Sri Lanka and Hong Kong.

Cadbury continued to grow its businesses overseas in steady incremental steps, tailoring products to local markets. Mars's attempt to venture into New Zealand in the early 1960s was quickly countered with a series of successful Cadbury countlines, culminating in Moro, which became so popular it was hard for Mars and Nestlé to gain ground. There was a clash when both Mars and Cadbury wanted to buy the leading Australian firm of MacRobertson in 1967. Cadbury succeeded and it wasn't long before Cadbury-MacRobertson secured 60 per cent of the Australian market.

The Quaker firm that a hundred years previously had had one traveller and had struggled to achieve national reach now stretched across the globe, but for Adrian it retained its human scale: 'I appointed all the people who ran our businesses outside this country. I knew them all, I visited, there was a link – if you like, we belonged to a family.'

ᗝᏕ❀Ꮥᗢ

Of all the markets in the world in the 1970s, the great prize was America. Once Forrest Mars had united the two arms of the Mars empire, he was determined to wrest supremacy in the American market from Hershey. He would not waste time on numerous lines, or fussy boxes of chocolates. He would rely on shifting volume of 3 Musketeers, Snickers, Milky Way and Mars Bars to reach his goal.

Unlike Mars, Hershey had failed to exploit global opportunities, and remained content to sell just in America. The conservatism of the Hershey establishment would play right into the hands of Forrest Mars. After a rapid rise in cocoa prices in the early 1970s, Hershey executives were looking for savings. Their first move was to slash the advertising budget. The Mars sales team, inspired by Forrest's near-religious fervour, went on a crusade to promote Mars products into every possible store at Hershey's expense. Mars was fast overtaking its long-standing rival.

In a surprising twist, Forrest Mars chose this very moment to walk away from the firm he had put so much into. In 1973, aged sixty-nine and at the peak of his achievement, he handed the reins to his sons, John and Forrest Junior. They continued the sales battle that he had begun, and the following year, in a blaze of triumphant headlines, Mars took Hershey's position as the number one confectioner in America.

The Cadbury team also had their eyes on America. 'The challenge was that the American market was 70 per cent owned by Mars and Hershey,' says Adrian Cadbury, who succeeded Lord Watkinson as chairman of Cadbury Schweppes in 1974.

He had already approached Hershey once to discuss a possible merger. For both companies, joining forces would bring considerable advantages. Hershey would gain an entrée into global markets, and

Cadbury a way into America. More important still for Adrian, there was a cultural and ethical fit between them that reached back to their early beginnings. For both, the business was about more than making money; both had close links with a loyal workforce and their local communities.

But as Milton and Kitty Hershey had always wished, their company was protected by the Hershey Trust, which managed the Hershey School and held approximately 80 per cent of the company's voting power. Nothing could happen without the support of the notoriously cautious Trust, and it rejected the proposed merger.

Undaunted, Cadbury Schweppes embarked on an alternative route into the USA, building a factory in Hazleton, Pennsylvania, and buying the Peter Paul Candy Company, which had 10 per cent of the American confectionery market, in 1978. Meanwhile, Adrian's drive to extend global reach was bringing results. By the end of the decade the export team was operating in 120 countries. The overseas strategy was co-ordinated down to the last detail. 'We produced international brand logos that set down the colour and the script,' said Dominic, 'and unified the whole presentation of the business across the globe.'

In 1984 Dominic was promoted to chief executive of Cadbury Schweppes, working alongside Adrian as chairman. They were the fourth generation of Cadbury brothers to be at the helm of the company. 'It really worked because I understood Adrian and he really understood me and we probably got the best out of each other as a result,' says Dominic. At the time the brothers were enjoying the runaway success in Britain of a new bar, Wispa. But before long they faced a predicament unlike anything their predecessors had ever encountered.

In 1987 the American company General Cinema, which owned a nationwide chain of movie theatres, acquired a massive 18 per cent of the shares of Cadbury Schweppes. The American leisure firm wanted to engineer a hostile takeover. 'They were trying to put us into play,' says Dominic. 'We went through uncertain times.' Over 160 years of Cadbury's independence was at stake.

General Cinema was able to take advantage of the fact that Cadbury's attempt to penetrate the American market did not go smoothly during the 1980s, which depressed the share price of Cadbury Schweppes. A major US launch of Wispa, Roses and other British favourites was stalling. With literally hundreds of lines to sell, wholesalers failed to give Cadbury's products an extra push to secure orders from retailers. Cadbury's chocolates were languishing in stock rooms as Cadbury US's profits tumbled.

On 22 July 1988, Cadbury US was sold for $300 million to its rival Hershey, which regained its lead in the American market from Mars. General Cinema retained its 18 per cent stake in Cadbury Schweppes. Noting the presence of American 'tanks on the lawn', the *Guardian* pointed out on 10 October 1990 that General Cinema 'did not intend to be passive shareholders'. Many believed that General Cinema itself would try to take over Cadbury, or would facilitate a takeover from a rival like Nestlé, and there were repeated rumours that a bid was imminent. General Cinema did not deny them.

It was an anxious time for Adrian and Dominic. But the brothers and the Cadbury Schweppes management team had a strategy to raise the share price to protect them from the American predator.

The year 1988 proved to be a critical one for the British chocolate industry. The most decisive change came in York.

The saga began to unfold on 13 April, when the Swiss-German firm of Jacobs Suchard made a 'dawn raid' on Rowntree, acquiring 15 per cent of the company's shares. Jacobs Suchard had already snapped up smaller European confectioners such as the Belgian Côte d'Or and the Dutch van Houten. Now it set its sights on Rowntree – but it was not the only one with eyes for the British firm. On 26 April, Nestlé's directors made a £2.1 billion hostile bid for outright ownership of the great Quaker firm.

Rowntree held the appealing prospect to the winner of a worldwide position in chocolate confectionery. Rowntree was the fourth largest chocolate-maker in the world, after Cadbury and the American firms of Mars and Hershey. Like Cadbury, Rowntree was well established in the old Commonwealth, and was exporting to over 130 countries. In addition to its merger with the British toffee manufacturers Mackintosh, it had also merged with the oldest French company, Chocolat Menier.

Not surprisingly, Nestlé's bid to buy Rowntree was greeted with outrage in Britain. British manufacturing had yet to feel the full force of globalisation, and people were appalled at the idea of a much-loved British chocolate icon falling into foreign hands. Was it fair, asked business leaders, for a Swiss multinational which effectively enjoyed immunity from takeover under Swiss law to freely acquire an important British company? There was a real fear that Nestlé would move production to cheaper plants overseas. And what about the different values and cultures of the two firms? Rowntree had over a century of tradition of being an integral part of York life.

In stark contrast to the high Quaker standards that Joseph Rowntree had once set for his firm, for many years Nestlé had been under attack for practices which some people regarded as unethical. At the heart of the question was the very issue about which Joseph Rowntree had felt so strongly: unprincipled advertising and promotion. This first came to light when the anti-poverty charity War on Want published *The Baby Killer* in 1974 – later retitled in one German reprint as *Nestlé Kills Babies*. These claimed that babies in the world's poorest countries, where poor sanitation and lack of clean water could spell a death sentence, were dying because mothers were inappropriately being encouraged to favour infant formula over breastfeeding. Nestlé successfully sued for libel, but the issue did not go away. Eventually the World Health Organisation developed international guidelines for the marketing of breast-milk substitutes, but allegations that Nestlé had breached them prompted further protests.

With opposition to the takeover mounting from all quarters, Rowntree's management rejected Nestlé's offer. At Bournville, chief executive Dominic Cadbury could see a potential solution. Cadbury would keep Rowntree British by taking it over itself – but only if local competition rules could be relaxed. Cadbury's management approached the Department of Trade and Industry. 'We said if you look at Cadbury and Rowntree's market share on a global basis – which is how you should look at it – there is not a competition problem.'

Nevertheless, Dominic was told that if Cadbury proceeded with the acquisition it would be referred to the Monopolies Commission: 'Civil servants in 1988 were not thinking about global market share, and did not see they were losing the chance to create a British world leader in chocolate. It was *the* great opportunity that went begging, but we didn't have a prayer of pulling it off with government thinking at the time.'

On 25 May, Margaret Thatcher's Conservative government announced that a foreign group could take over Rowntree. The next day Jacobs Suchard topped Nestlé's bid with an offer of £2.3 billion. In mid-June Nestlé raised its bid to £2.5 billion. The Rowntree Trusts, which had once had a 51 per cent stake in the company, now had less than 9 per cent. This was partly due to their having diversified their share portfolio, but also because their holdings had been diluted following successive share issues. The decision about whether or not to accept Nestlé's bid was down to Rowntree's shareholders – a diverse group – and they voted in favour. Overnight, Nestlé became one of the world's top four chocolate confectionery firms, with a large share of the British market, and a famous Quaker company had had a taste of unfettered shareholder capitalism.

For Dominic Cadbury, the idea that Nestlé should be allowed to buy Rowntree's, while Cadbury was not, seemed 'ridiculous': 'That was a big fork in the road,' he remembers. 'We as a country have shot ourselves in the foot here. Had we been more forward-thinking about *global* market share we would have pushed Cadbury and Rowntree together.'

Since the takeover, staff numbers at Joseph Rowntree's great factory at Haxby Road have declined to 1,600, and the manufacture of famous brands like Smarties has been moved overseas. Even the Rowntree name has been discreetly dropped from the packaging on many brands.

Nestlé was the world's largest food company, and now had over 7 per cent of the confectionery market. Today its presence is felt in almost every country in the world, and its company literature considers developments in food in the context of feeding the global population. Questions about its size and market dominance seem to dissipate under the sheer might of the company itself.

General Cinema still held almost a fifth of Cadbury Schweppes's shares, but the Cadbury brothers were strengthening the company's independence by building on their global lead in confectionery and drinks.

Sir Adrian Cadbury, one of the firm's longest-serving chairmen.

The Cadbury team could not grow the chocolate division of the business through mergers with its natural partners, Hershey or Rowntree's, but it was able to expand its confectionery division by acquiring strong brands such as Lion confectionery and the Bassett and then the Trebor groups. During the 1980s Dominic Cadbury streamlined chocolate production. State-of-the-art technology was introduced at Somerdale and Bournville, and the number of brands manufactured was reduced from seventy-eight to thirty-three. The East Cocoa Block at Bournville was converted into a visitors' centre, Cadbury World, which quickly established itself as one of the most popular tourist attractions in Britain. Dominic was keen to find ways to make the drinks side of the business more dynamic as well, by creating a new bottling company with Coke called Coca-Cola Schweppes and acquiring American soft-drinks brands such as Canada Dry. As they refocused the business, the Cadbury brothers saw the share price recover.

Despite the efforts of General Cinema, no bidder came forward. At no stage did Adrian or Dominic agree to meet General Cinema for any negotiation over the future of the firm. Eventually after three worrying years, General Cinema sold its shares. 'It was hugely satisfying to see this threat to the company disappear,' Dominic says. 'One of the best moments of my life.'

In the years following Adrian Cadbury's retirement in 1989,

Cadbury Schweppes continued its global expansion. The collapse of the Soviet Union unlocked a vast new market. Both Mars and Cadbury built chocolate factories in Russia, and by 1995 they were in Beijing, tackling the largest market of all. Under Dominic's era as chairman the acquisitions continued, notably in drinks, first Dr Pepper and 7 UP, and later Snapple. 'Renowned as a marketing whiz', reported Andrew Davidson in August 1997 in *Management Today*, Dominic had helped to position the company as 'a truly global multinational' that could compete with Coke and Pepsi 'with astonishing audacity'.

When Dominic stepped down in 2000, after six years as chairman, for the first time in the firm's 170-year history there was no member of the Cadbury family on the board. At this point less than 1 per cent of Cadbury Schweppes shares were in family hands. Over the years the shares held by the Cadbury benevolent trusts had also declined, as their financial advisers had recommended that they diversify their holdings. 'That is simply a prudent decision by the trusts,' Adrian says, 'and in my view the wealth which the company produced for the family continues to be used for the causes that the family holds dear.'

The shareholders in Cadbury Schweppes, however, were increasingly made up of investors who had no direct personal links to the business and its values, and whose priorities were purely to maximise profit. The Quaker voice no longer held sway in the boardroom, but could it be heard anywhere in the modern drinks and confectionery giant?

Dominic argues that it could, proudly citing his brother Adrian's role in developing Britain's code of corporate governance: 'I think you can say the Quaker DNA has shone through the Cadbury company in terms of the work that the former Cadbury chairman has done to help develop the first code of best practice.' Under Adrian Cadbury's chairmanship in 1991, the Committee on Financial Aspects of Corporate Governance set a code of best practice on critical issues, among them honest disclosure, excessive executive pay and balancing short-term and long-term interests. The code was the basis for effecting widespread reform of corporate governance. 'Eleven years and twenty-eight countries later,' wrote Simon Caulkin in the *Observer* on 27 October 2002, 'Cadbury is the elder statesman of the corporate governance movement and Britain the corporate governance capital of the

world.' Although Quaker values were not explicitly mentioned in the
Cadbury Code, for Adrian they were crucial. The aim of the code,
he said, was to bring 'greater transparency, honesty, simplicity and
integrity to the process of running a company'.

Even as a global corporation, the company tried to stay faithful to
its Quaker heritage, says Todd Stitzer, Cadbury's chief strategy officer
in 2000. A Harvard-educated lawyer, Stitzer chose to stay with the
British chocolate company all his working life: 'I admired the culture.
It's the appeal of the head-heart relationship that existed within the
business. It mattered hugely to me.'

❧❀❧

In the new millennium the confectionery industry faced a fresh wave
of consolidation. The Hershey Company, so long protected by Milton
Hershey's will, became the focus of unwelcome attention once more.
Just as the Cadbury trusts had diversified their shareholdings, directors
of the Hershey Trust also began to question whether the Hershey
School would be better protected with a more diverse source of
income. 'They literally decided that they would auction the Hershey
Company,' recalls Todd Stitzer, 'and they had a line of potential buyers.'
Wrigley, Cadbury and Nestlé began separate negotiations with the
Hershey Trust.

When news of the secret meetings was leaked there was an immediate outcry. The Hershey community felt betrayed. Posters were
printed warning 'Wait 'til Mr Hershey finds out!' Residents and workers
paraded down Chocolate Avenue reminding anyone who would listen
of Milton Hershey's proud heritage. The Pennsylvania state attorney
general, Mike Fisher, was deluged with complaints, and mounted a
legal challenge to any sale. On 3 September 2002 the case came
before the court, and the judge ruled that no sale could happen
without his approval. Hershey remained independent.

Todd Stitzer, promoted to Cadbury's chief executive in 2003, was
charged with growing their confectionery business. There were very
few public chocolate companies left to acquire. After the failure to
secure a deal with Hershey, he changed strategy, buying sugar confectionery and gum businesses in different parts of the world. The acquisition for $4.2 billion of the New Jersey-based Adams, the world's

second largest chewing gum concern, turned Cadbury Schweppes overnight into the world's largest confectionery giant.

The company was growing so fast that Stitzer was concerned that it might lose touch with its core values. 'We consciously said in 2003 that we were going to magnify and modernise the George Cadbury principle that "Doing good is good for business," ' he said. Even as number one in world confectionery, he wanted to try to embed Quaker values in the business. The Quaker founders weren't just philanthropists, he argued, 'they were *principled* capitalists'. They saw themselves as long-term stewards, committed to all the stakeholders in the business – the staff and the wider community – not just to making profits for themselves.

Stitzer acknowledges that modern capitalism can be destructive, 'a one-way relationship in thrall to profit margins and shareholder returns'. Equally true, however, is that over-regulated capitalism can constrain creativity and innovation. He believes there is a middle way, which he terms 'principled capitalism', where business leaders build long-term value for shareholders in a 'socially responsible way'.

To embed the values of the company in the business plan in a way that could be applied in all countries, Stitzer set a target of 1 per cent of pre-tax profit to be committed to programmes that benefited the communities in which the business operated – a target that was consistently exceeded. Soon other industry-leading initiatives were launched, such as 'Purple goes Green', a commitment to reduce absolute carbon emissions across the company by 50 per cent by 2020. Ambitious plans were also in progress to ensure ethical sourcing of cocoa and to provide help to farmers.

This was all the more urgent, since in the new millennium worrying reports were emerging of child-labour abuses in West Africa, particularly in Ivory Coast. Although Cadbury bought its cocoa from the neighbouring country of Ghana, it was becoming apparent that the problem existed throughout the region. How could this be reconciled with principled capitalism? Cadbury was already involved in an extensive well-building programme across Ghana to provide rural villages with fresh water, but Stitzer's team wanted to do more. The company joined forces with the United Nations, Anti-Slavery International, World Vision, Care and VSO to create a 'Cocoa Partnership'. Cadbury would make a £45 million commitment over ten years to enhance the lives of cocoa farmers in Ghana, India and the Caribbean by investing in their farms while also building schools and infrastructure in rural districts.

The Cadbury team also began to collaborate with the Fairtrade Foundation to tackle injustices of trade and to help farmers out of poverty. Fairtrade ensures a guaranteed minimum price of $1,600 per ton to cocoa producers, even if world cocoa prices fall below this level. In addition, however high the price of cocoa rises, Fairtrade guarantees a $150 premium per ton on top to help farmers to develop their businesses and communities. Although even Fairtrade cannot rule out the possibility of child-labour abuses on its farms, it provides better traceability and more direct contact with cocoa producers, which makes it possible to identify and deal with bad practice.

Stitzer knew that if he could convert a leading brand like Dairy Milk to Fairtrade, it would be an industry-leading move that would triple the amount of Fairtrade cocoa exported from Ghana at a stroke. Many shareholders investing for the long term supported the strategies Cadbury was adopting. But as Stitzer wrestled to align the world's largest confectionery company with the values of principled capitalism, he faced a fresh challenge – one that put the future direction of Cadbury Schweppes in jeopardy.

It began with a small change to the share register.

In 2007 the American billionaire Nelson Peltz bought 3 per cent of Cadbury Schweppes through his hedge fund, Trian Investments. 'Nelson Peltz is a force of nature,' declared Shawn Tully in *Fortune* on 19 March 2007. Even at sixty-four, he remained 'relentlessly competitive'. As an activist investor, Peltz could see a way to bring short-term returns to shareholders: a separation of Cadbury from Schweppes.

Separating the drinks and confectionery businesses had already been under discussion by the Cadbury Schweppes board as a possible move to release value to shareholders. The combined company was worth around £12 billion, while separately the drinks arm was esti-mated at £7–8 billion, and confectionery at £9 billion. But there was a catch – a vital one for those who wanted Cadbury Schweppes to remain a British independent. The sheer size of the drinks and confectionery giant meant that a potential buyer would find it hard to raise enough money for a takeover. It had an awkward structure too; any buyer would almost certainly choose to break up the company. But if the drinks half was spun off, Cadbury's confectionery could become a tempting takeover target. Some on the Cadbury Schweppes board argued that it would only make sense to proceed with the sale of the drinks division if another confectionery acquisition was lined up. With no such deal in place, whenever asked in public by investors about a possible sale of the drinks division, Stitzer and the Cadbury Schweppes chairman John Sunderland said it was not going to happen.

Behind the scenes, Stitzer drove yet another initiative to unite Cadbury and Hershey in 2007. This time they got close. There was a measure of agreement between the Hershey Trust and Cadbury, but the management of Hershey Foods persuaded the Trust not to go through with it. Meanwhile, Nelson Peltz ramped up the pressure, publicly agitating to split Cadbury Schweppes in two and prompting a debate in the financial press about the dawn of a new age of 'offensive share ownership'. The increase in the number of international institutional investors whose loyalties did not lie with British management made it easier for Peltz to influence strategy.

Stitzer concedes that there were differing views among the board. 'There were board members who would rather have seen the businesses stay together, but that was a very difficult thing to do in the face of significant pressure from a large number of shareholders.' Facing wide-ranging pressure to enhance shareholder value, in the spring of 2007 the board of Cadbury Schweppes reached a unanimous decision to split the company. However, just at that moment the global credit crunch began to take hold. As the financial crisis escalated, the sale of the drinks business, now called Dr Pepper Snapple Group, collapsed when the private equity buyers could not raise the anticipated £7–8 billion price tag.

Nelson Peltz was determined to unlock the value in the two companies. If a sale of the drinks division wasn't possible, he wanted to refashion the transaction as a de-merger. Above all, he wanted higher margins. On 18 December 2007 he sent an open letter threatening the Cadbury Schweppes board. Management 'had nowhere to hide', he said, and its credibility with shareholders was 'very low'. He demanded that the board appoint several new directors and immediately name a new chairman to replace Sir John Sunderland, who was due to retire. Two months later, the board announced that Roger Carr would become the chairman of Cadbury's confectionery when the de-merger was completed. Carr had joined the board of Cadbury Schweppes in 2001. He was well known in the City as chairman of Centrica, and had held board positions at other leading FTSE 100 companies.

'Nelson literally lobbied our major shareholders to remove the management of Cadbury,' Carr says. 'He put the case that the Cadbury management were completely inept, why they should be removed . . . and why he or his representatives should be installed to extract value from the company.' Carr too went round to see all the large shareholders, pointing out that the existing management under Todd Stitzer had bought Adams and transformed the business. 'So don't let's suddenly decide we want to throw out the chief executive and the finance director,' he said.

Stitzer accepts that his job was threatened, but says that this was not connected to the beverages de-merger. 'It was related to the margin generation and the management's ability to control costs,' he says. He acknowledges that in 2006 Cadbury's profits were hit by a salmonella scare and an accountancy fraud in a factory in Nigeria. 'We didn't deliver the margin we said that year, and that caused a stir among some of the shareholders, saying, you know, "This is the gang that couldn't shoot straight." ' Stitzer's frustration is palpable: 'People were – as short-term shareholders can often be – just focused on "Give us cash, give us margin." And we were trying to do it the *right* way.'

In the spring of 2008 the costs of the de-merger soared to an estimated £1 billion – a staggering 10 per cent of the value of Cadbury Schweppes. Some investors began to protest that this was too high a price for 'a bit of focusing'. Nelson Peltz, who now owned 4.5 per cent of Cadbury, continued to agitate for a split, and the company pressed ahead with the de-merger – ironically at the very time that there was further consolidation elsewhere in the confectionery industry. In May 2008, Mars merged with the chewing-gum giant Wrigley. The £11.6 billion deal toppled Cadbury from its number-one slot. Mars-Wrigley became the world's largest confectionery company.

By contrast, was Cadbury – now shorn of its drinks division – a potential sitting duck?

Roger Carr does not think so. 'The de-merger was the right thing to do,' he says. 'The role of companies is not to remain independent at all costs but to create value for those who own them . . . Certainly by being smaller you become more vulnerable to takeover, but there is nothing wrong with that – that is one way that value is created.'

This raises the question: shareholder value over what period? If the board prioritises the creation of short-term value for shareholders, where does that leave the wider interests of the company: the work-force, the cocoa growers, investment for the future and the creation of long-term value? And if short-term value comes at the cost of breaking up a company, there may be long-term consequences that include sacrificing the company's independence.

It wasn't long before the company had an unwelcome approach. 'I was at the airport coming back from Lisbon in late August 2009,' Carr recalls. 'There was a voicemail on my mobile saying, "I'm Irene Rosenfeld. I'm in the UK next week and wouldn't mind coming and having a cup of coffee . . . " '

Irene Rosenfeld was the chairman of Kraft Foods, America's largest food giant.

18

They'd Sell for 20p

Irene Rosenfeld had been watching Cadbury, waiting for the right moment. Potentially even more valuable to her than Cadbury's large slice of the British market was its position in global markets – especially in fast-growing developing countries. Cadbury was the foremost chocolate brand in Africa, Australia, India and numerous other countries such as Malaysia, Singapore and New Zealand, and it was developing a presence in Russia and China. Now separated from the Schweppes division, Cadbury's confectionery business was valued at over £10 billion, with annual sales of £5 billion. For the American food giant, which was five times larger, Cadbury was a tempting target.

After working her way up the corporate ladder, Irene Rosenfeld had become chief executive of Kraft in 2006, and chairman a year later, at a critical point in the firm's history. Much of the transformation of Kraft happened between 1988 and 2007 under the wing of America's leading tobacco corporation, Philip Morris. Kraft gained full independence from the tobacco firm in 2007, emerging as the second largest food company in the world and a fitting challenger to Nestlé. Kraft operates 168 factories worldwide, has 98,000 employees, and generates annual sales of over £26 billion, while Nestlé, the number one food company in the world, has five hundred factories and 250,000 employees, with annual sales of over £72 billion.

The story of Kraft Foods Inc. began in 1903, when a twenty-nine-year-old Canadian entrepreneur, James Lewis Kraft (known as J.L. Kraft), opened a wholesale cheese business in Chicago. Originally from Stevensville on the shores of Lake Erie in Ontario, he struggled to make ends meet, and was down to his last $65, which he invested in a horse called Paddy and a rented wagon. His idea was to buy cheese in bulk from wholesalers in South Water Street in Chicago and resell it to individual grocery stores across town, but his plan did not go smoothly. 'Paddy and I were equally discouraged,' he said. 'My small capital exhausted . . . I was a failure.'

Kraft cheese wagon c.1921

But he persevered, and gradually won the confidence of the local grocers. After a year he was able to invest in more horses and carts. As the business prospered, his brothers joined him, and by 1913 they were selling thirty different types of cheese. The turning point came with the onset of the First World War. J.L. Kraft was particularly interested in how to extend the shelf life of cheese by processing it. He found that when cheese was heated with emulsifiers, whey and other dairy products, it did not need refrigeration and could travel long distances – exactly what the US Army needed. Kraft provided six million pounds of tinned and processed cheese to the military during the war, and did not look back.

By 1930, J.L. Kraft had captured 40 per cent of the American cheese market and was operating on three continents. He bought other companies, notably the Phenix Cheese Company, makers of Philadelphia cream cheese. He expanded his range of convenience foods in the interwar years, introducing Velveeta cheese spread, Miracle Whip salad dressing and boxed Macaroni and Cheese Dinner. Kraft had another success on his hands when he pioneered the first ready-sliced processed cheese in 1950; it was an instant hit when combined with that all-American icon, the hamburger. 'It was truly revolutionary,' says Kraft Food's archivist, Becky Tousey. 'They had to have in-store demonstrations before customers could believe the slices would easily separate.' Later Kraft hit on the idea of wrapping each slice of cheese in cellophane for convenience. By the time J.L. Kraft died in 1953, his company had become a household name in America.

The dramatic transformation of Kraft from a successful American firm to the world's second largest food company started in 1988 when it was bought by Philip Morris. The tobacco giant had very good reasons to diversify its business. In 1954, in one of the earliest cases of tobacco litigation, a Missouri smoker who had lost his larynx to cancer, filed suit against Philip Morris. The tobacco company won the case in 1962, but the problem did not go away. As evidence suggesting a link between cigarette smoking and cancer mounted, so did the costs of litigation. In 1988, after a long-running court case, the judge said he found evidence of a conspiracy by three tobacco companies – including Philip Morris – that was 'vast in its scope, devious in its purpose, and devastating in its results'.

That very year, Philip Morris diversified further into food, buying Kraft for $12.9 billion in one of the largest non-oil corporate takeovers in US history. The tobacco company had already bought General Foods for $5.6 billion, a large food concern that owned many famous brands including Walter Baker, America's oldest chocolate firm. Irene Rosenfeld had begun her own career at General Foods, progressing through various managerial roles, becoming one of the company's first two female general managers as she did so.

The sheer scale of the mergers that followed beggars belief. In 1989 General Foods merged with Kraft, and soon acquired Jacobs Suchard, which brought with it Suchard chocolates and the Tobler company, makers of Toblerone. In 1993 Kraft General Foods bought the historic British chocolate confectioner Terry of York, acquiring Terry's Chocolate Orange and other much-loved treats. In 2000 Philip Morris bought Nabisco Holdings, America's number-one biscuit-maker, for a staggering $18.9 billion, and merged the company with its Kraft division. Philip Morris, apart from being one of the largest cigarette companies in the world, was rapidly becoming a colossal food concern.

Then, in 2000, Philip Morris and R.J. Reynolds were ordered to pay $20 million to a smoker who was dying of lung cancer. It was the first ruling to hold cigarette-makers responsible for the health of people who took up smoking in spite of the package's compulsory warning labels. On 7 June 2001, Philip Morris was ordered to pay $3 billion to a smoker with terminal cancer – a record-breaking individual damage award against a cigarette-maker. One week later Philip Morris raised $8.7 billion by selling 16 per cent of Kraft Foods.

Kraft Foods was now listed on the New York Stock Exchange, but it was still principally owned by Philip Morris. In 2003 Philip Morris changed its name to Altria, which still owned the majority of Kraft's stock. Although the judgements against the tobacco corporation were reduced on appeal – the $3 billion damage award was lowered to $82 million in March 2006 – more lawsuits continued to be filed against the company. The following January, the Altria group voted to spin off all remaining shares of Kraft Foods.

Kraft Foods finally became independent of tobacco on 30 March 2007. Its holdings included Maxwell House coffee, Philadelphia Cream Cheese, Oscar Meyer hot dogs, Nabisco biscuits and snacks, Dairylea, Terry's chocolates and Kraft cheeses. With its complex history, Kraft, declared London's *Evening Standard* on 9 September 2009, was 'a creature of Wall Street, an assemblage of businesses . . . that were stitched together by Philip Morris during the merger mayhem of the 1980s and 1990s'.

Irene Rosenfeld was appointed chairman of Kraft Foods in March 2007. She knew that despite the company's phenomenal size, many of its brands were established in developed markets, yielding low growth of around 4 per cent to shareholders. With Cadbury's stronger footprint in faster-growing developing markets, she saw the potential to raise this figure to 5 per cent growth.

On 26 August 2009 she flew to Luton in the Kraft company jet and made her way to the Ritz Hotel in London. At 9.30 the following morning she went to see the Cadbury chairman, Roger Carr. The meeting was discreetly held in his office in Centrica's headquarters in Burlington Lane. Carr has a reputation as a 'City grandee', according to Andrew Davidson of the *Sunday Times*. 'A hard man to read,' says Davidson, 'and as cautious and leathery as an old tortoise.'

The meeting did not take long. Carr remembers that after about three minutes of pleasantries, 'She said, "You know, I have this great idea that we should buy you." ' She told him her plan was to offer a cash and shares bid for Cadbury worth £10.2 billion. Carr describes Rosenfeld as 'clinical, distant, and quite hostile. She showed no natural warmth. Her body language was driven and intense – certainly not relaxed and engaging.'

Poker-faced, Carr did not hesitate, but replied, 'Well, first of all, this is something I will want to discuss with the board, and secondly, Cadbury is a very good business, it's doing very well as an independent, and certainly doesn't need Kraft.'

After a brisk exchange, Rosenfeld said she would courier round a letter that afternoon, and asked for his response by Wednesday.

'We'll give you a response when we think it is appropriate,' Carr responded. He walked her to the lift, 'and off she went'. The meeting, he recalled afterwards, did not last more than fifteen minutes.

Later that day, Rosenfeld's letter arrived. 'I very much enjoyed meeting you this morning,' she began with pro forma courtesy. Her letter set out a textbook case for globalisation. Kraft's purchase of Cadbury would be the logical next step as 'we shape the company into a more global, higher growth and higher margin entity'. The new company would have $50 billion in revenues each year, and 'scale in key developing markets such as India, Mexico, Brazil, China, and Russia'. The 'strong presence in instant consumption channels in both developed and developing markets' would expand the reach of the business and provide 'potential for meaningful revenue synergies over time'. There was also the possibility of savings. In a subsequent letter, Rosenfeld explained how the acquisition would save $300 million in economies of scale in manufacturing, $200 million in administration, and $125 in marketing and media.

An emergency Cadbury's board meeting was held at the bankers Goldman Sachs's offices on Fleet Street. 'The mood was, "We will not allow these people to steal this company," ' recalls Carr. 'Everyone had utter resolve around the board table to resist this.' Carr drafted a letter rejecting the offer. The plan to bring Cadbury into Kraft's 'low-growth conglomerate business', he said disparagingly, was 'an unappealing and unattractive prospect'.

Rosenfeld's next move was to publish the letter she had sent to Carr, initiating what is known in the trade as a 'bear hug'. When letters of intent are made public, says Carr, 'the predator can distress and disturb the prey whilst alerting the market to the potential for an exciting bout and quick financial gain'. The audience is in no doubt 'that a showdown is inevitable'.

The showdown soon began. The British press was hostile. Rosenfeld's bid was little more than 'brazen imperial ambition', declared the *Evening Standard* on 9 September. Kraft was caught in a static American market that 'rises and falls with the waistline of Joe the plumber'. Many pointed to the fate of Terry, the cherished British chocolatier, under Kraft's stewardship. Chocolate production had been moved to Eastern Europe, and the historic factory in York closed in 2005. Felicity Loudon, George Cadbury's great-grand-daughter, condemned Kraft as a 'plastic cheese company', and voiced fears that it could asset-strip 'the jewel in the crown'. The *Sunday Times* summed up the force of the British opposition: 'Cadbury Gives it Both Barrels' read the headline, with an image of Todd Stitzer blasting the US predator.

Irene Rosenfeld did not waver. In what was seen by many as an unnecessarily hostile move, she took her offer straight to Cadbury's shareholders. The prospect of a takeover prompted a shopping frenzy. Hedge funds and other short-term investors piled into Cadbury as the share price soared. What, they wanted to know, would maximise their profits?

News of Kraft's proposed takeover sent Cadbury shares soaring. City sharks and other predators began circling around the chocolate prey looking for a quick kill. Bankers and accountants gutted the balance sheet. There was talk of carving up Cadbury's assets. Could parts of the company be acquired for a knockdown price? Investors had their eyes on the most profitable brands, Dairy Milk and Trident gum. If the confectionery industry was about to be massively realigned, no one wanted to be left on the sidelines.

Hershey executives woke up to the threat that their company could be left behind in a new world of behemoths like Kraft-Cadbury, Mars-Wrigley and Nestlé, and arranged a series of meetings with Cadbury. Among the many issues on the table were the questions of how the Hershey Trust could keep control with a massively reduced shareholding in a combined company, and how Cadbury shareholders would benefit from a merger. Also, Hershey was half the size of Cadbury, so how could it afford the acquisition? If the two companies could resolve these and other issues, Carr was confident that a merger would be 'a wonderful outcome'.

Cadbury was under siege. Hedge funds, which had previously owned 5 per cent of Cadbury shares, bought 20 per cent in a matter of weeks, and the share price rose from around £5 to above £8, severely testing the loyalty of long-term investors and opening the door for hedge funds to continue to pile in. Carr points out that British institutional investors had already turned their backs on Cadbury. At the start of the bidding process, only 28 per cent of Cadbury shares was British-owned, as opposed to 50 per cent owned by Americans. Now these American investors were tempted to cash in as the share price soared.

With the ownership of Cadbury changing fast, further destabilising the company, on 21 September management asked the UK's Panel on Takeovers and Mergers to give Kraft a 'put up or shut up' deadline, which required Kraft to make a formal offer or walk away.

As shareholders rushed to evaluate their options, the billionaire investor Warren Buffett, who owned 9 per cent of Kraft, spoke out. The 'Sage of Omaha', as he is known to admiring investors, is one of the richest men in the world, a position earned after a lifetime of walking with care through inflammatory markets. He urged Kraft not to overpay for the British chocolate firm, and it looked as though the Kraft management was listening.

On the day of the Takeover Panel deadline, Rosenfeld made a formal bid at the same price as her earlier offer. But because Kraft's share value had declined slightly, the bid amount was now worth less: £9.8 billion, or £7.17 per share. Roger Carr dismissed the offer as 'derisory'. It was beginning to look as if Kraft could not afford Cadbury.

On 18 November, news broke that the Italian firm of Ferrero Rocher was joining the chocolate wars. Ferrero, the family company behind Nutella and Ferrero Rocher chocolates, was even smaller than Hershey. Could it possibly join forces with Hershey to make a combined bid for Cadbury? Confirmation that the Hershey Trust was reviewing a possible bid for Cadbury fuelled excitement that Kraft's bid would be topped. Then Nestlé revealed it was considering joining the bidding war. Just two months after Kraft's opening salvo, amid speculation that Hershey – or someone else – might produce an $18 billion bid, Cadbury's shares soared by 40 per cent.

On 14 December, Cadbury's management issued a bullish defence. Stitzer promised greater returns to shareholders if the company remained independent, with the prospect of 5 per cent annual growth and double-digit dividends. These forecasts were backed up by the company's third-quarter results. Cadbury's sales were better than expected, in contrast to Kraft, which had to cut its 2009 sales forecast.

After a cold and snowy Christmas in London, in early January 2010 Kraft revealed that only 1.5 per cent of Cadbury shareholders had accepted Kraft's bid. But Rosenfeld was steadily moving closer to her goal. She sold Kraft's North American frozen pizza business to Nestlé for $3.7 billion. Despite opposition from Warren Buffett, the sale gave her extra cash for the Cadbury bid and made it less likely that Nestlé would join Hershey or Ferrero in an attempt to outbid Kraft. In addition, according to Adam Leyland in the *Grocer* magazine on 23 January 2010, Buffett's warnings against Kraft overpaying for Cadbury 'helped Kraft shares recover, upping the value of the bid' and serving to 'low-ball expectations'.

There was one usually vocal American investor who 'remained uncharacteristically quiet', wrote Leyland. Nelson Peltz, who had agitated for the de-merger of Cadbury Schweppes in 2007, had at the same time taken a significant position in Kraft. Moreover, 'Peltz secured a two-year deal with Kraft management not to publicly criticise the company in exchange for two independent director appointments on Kraft's board,' says Leyland. This 'gagging deal' expired during Kraft's bidding process. Yet Peltz remained quiet. 'By this time he had also, intriguingly, sold the majority of his shares in Kraft,' says Leyland. 'Anyone looking for a silent player behind the scenes driving this deal should look no further than Peltz.'

The week before Kraft's deadline to make a final offer on 19 January, speculation rose that Hershey was about to mount a solo bid. There were anxious meetings in London hotels. 'Until the very, very end, Hershey was still trying to find a way to increase the consideration in a manner they could finance appropriately,' says Stitzer. 'They couldn't get to a place where they needed to be.' Carr was more blunt. The Hershey Company, he said later, was 'paralysed by internal conflicts of opinion'.

By this time, short-term investors such as hedge funds owned as much as 31 per cent of Cadbury. Carr was talking regularly to the shareholders. 'Some of the hedge funds said to me, "We've bought at £7.80 – with 20p in five weeks of ownership – we'll sell for £8." The ones that came in later, maybe they bought at £8. They'd sell for the same 20p – but the clearing price became £8.20.' British institutions still held some 28 per cent of Cadbury, and some of them wanted above £8.50 – but they were in a minority. 'A lot of the American owners said they would sell in the £8.20 to £8.30 zone for sure.'

In London, Warren Buffett's warnings not to overpay for Cadbury had been widely reported, and many investors believed that Kraft could not afford to increase its offer significantly. As the deadline approached, Cadbury shares began to fall on the expectation that the bid might fail. There was talk that Rosenfeld might be obliged to make an embarrassing retreat. Cadbury might yet get away.

That weekend Rosenfeld returned to London and settled into her suite at the Connaught Hotel in Mayfair. At seven o'clock on Sunday evening she rang Roger Carr and arranged a meeting for the next morning in a private room at the Lanesborough Hotel. It was Rosenfeld's final chance to win Carr's approval for the bid, which would make the takeover far more straightforward.

Carr remembers vividly how the meeting started: 'She began by saying, "We've listened to you, we've listened to your shareholders, we know we have to pay more money, and I'm going to offer you £8.30." ' The minute she said that, 'I knew we'd lost,' he said. 'I knew the business was sold in the real world.'

Carr left to consult the other members of the Cadbury board. Having spoken to both shareholders and advisers, the board believed that if Rosenfeld had gone to the market the following day and offered £8.30, she would have secured more than 50 per cent of the shares immediately. She only needed 50.01 per cent.

'I knew she'd got it,' said Carr. 'My job from that point on became to get as much value as I could. The most important thing was to get it from £8.30 to £8.50, which was worth nearly another half a billion dollars for shareholders.'

But did Carr and the board capitulate too soon? 'By playing the heritage card so strongly in their defence against Kraft', Alex Brummer would observe in the *Daily Mail* on 2 February, they raised the hopes of all stakeholders 'that this was a genuine defence aimed at keeping independence rather than a bluff aimed at getting the price up'. Those in favour of preserving Cadbury's independence were left to wonder whether the board could have seen off the bid at £8.30 had they stood firm.

Carr doesn't accept this. 'We resisted the Union Jack defence and focused on value. I fought for the shareholders. I'm paid by the shareholders and I delivered huge value for the shareholders with the board – that is my responsibility.'

Later in the day on 18 January, Carr and Rosenfeld met again at the Lanesborough. After a series of meetings, Rosenfeld finally offered £8.40 with a 10p dividend once the offer had been unconditionally accepted. In essence, she was offering £8.50 per share. 'The board's view was that we had achieved a good price for the business,' Carr says, and they were prepared to recommend the bid.

It was dark as Irene Rosenfeld and Roger Carr made their way across Mayfair to Kraft's advisers and bankers at Lazard on Stratton Street. They were joined by Cadbury's advisers from Goldman Sachs. 'The transaction was secured at around 9 p.m.,' says Carr. Kraft's PR people asked for a photograph of Rosenfeld shaking hands with Carr. 'I said no, because I had never changed my position that I did not want to sell the business to Kraft,' Carr recalls. All the same, he felt he had done his job: he had secured 'tomorrow's price today'. But for him, this was not a moment for a toast: doing 'the right thing may personally leave you feeling sad and hollow'. Todd Stitzer too felt 'unspeakably sad'. At five o'clock the following morning he woke Sir Adrian Cadbury with the news, anxious to reach him before the story broke in the media.

The press was waiting for a statement. 'The board of Cadbury unanimously recommends Cadbury shareholders to accept the terms of the final offer,' said a weary-looking Stitzer. 'The deal represents good value for Cadbury shareholders,' assured Carr. The press was quick to point out that among those shareholders poised to benefit from the takeover was Stitzer himself, who reputedly walked away with an estimated £17 million in shares and options. 'The strange paradox of this is that investing in the company I believe in, in the end actually was of benefit to me,' he said later. 'I wasn't in any way seeking an early end to my employment.'

According to Carr, most shareholders were 'very pleased with the price achieved'. Bankers and advisers in London and New York also benefited to the tune of £400 million. Yet some shareholders had a muted response. Legal and General Investment Management, which held 5 per cent of Cadbury's shares, was 'disappointed' that the price did not reflect the company's true value.

Others were shell-shocked by the news. Felicity Loudon saw the outcome as 'a horror story', and urged people to voice their opposition to their MP and the shareholders. Many British consumers were equally angry and outraged. 'There should be a national boycott of Kraft-Cadbury,' urged one website, and campaigns were launched to save the Curly Wurly and other beloved brands. Some financial analysts thought the firm had been sold down the river. The two grandsons of George Cadbury, Sir Adrian and Sir Dominic, described the news quite simply as 'a tragedy'.

A smiling Irene Rosenfeld now stood at the helm of the world's number-one confectionery super-giant. Cadbury-Kraft combined was a global powerhouse, with worldwide sales of £37 billion.

Mars and Wrigley had been knocked into second place.

19

Gone. And it was so Easy

The hotly contested takeover of Cadbury was one of the largest business deals in British history. For many it was just part of an ongoing process that has brought substantial benefits worldwide. Globalisation has helped to lift millions out of poverty, and enables a wider range of products to be sold around the world at cheaper prices than before. In the chocolate industry, if the rationale that Kraft used to persuade Cadbury shareholders is correct and the projected synergies between the two companies are realised, Cadbury will become a leaner and more efficient organisation, and Kraft will sell more overseas, creating opportunities for all Kraft employees. Investors will enjoy higher profits, and confectionery and other goods will be produced ever more cheaply – at least, that is the theory.

Sir Adrian and Sir Dominic Cadbury warned in a letter on 20 January 2010 to the *Daily Telegraph* that a high percentage of takeovers do not live up to the bidder's claims. Kraft's record to date, they wrote, was of 'underperformance and, in the case of Terry, of failing in their stewardship of the company they had acquired'. The two former Cadbury chairmen pointed out that the value of a company reflects its reputation built up over generations, and consumers' trust that the company and its brands are one. 'Cut the tie and submerge the brands in a larger entity,' they wrote, 'and both present and future value will be lost.'

The difficulty of maintaining the culture of a company that is taken over is linked to the new owner's debt load. In this case, Kraft raised an estimated £7 billion to fund the takeover, increasing its total debt to a reported £18 billion. This staggering amount of debt has generated fears that Cadbury will be asset stripped to service it. Despite Kraft's assurances that this will not happen, the premise of the takeover acknowledged annual efficiency savings of £412 million. 'Nobody knows whether or not they can achieve it,' says Sir Dominic, pointing out that Kraft was not obliged to show how this – or synergies with Cadbury producing savings of £650 million – would be realised. So it is hardly surprising that the very stakeholders whose lives George Cadbury and the pioneers were at such pains to enrich are likely to lose out sooner or later – starting with the employees.

Sir Adrian and Sir Dominic appealed to Kraft on 20 January that having 'accepted a duty to those working for the company', it should 'live up to that responsibility'. Unite, the trade union representing Cadbury's employees, is concerned that up to 10,000 jobs in Cadbury worldwide could be at long-term risk. With Kraft under pressure to meet debt repayments and achieve its profits forecast, the union fears that the company will be more likely to cut jobs in Britain rather than on its home turf in the USA. Unite claims that over the last ten years Kraft has shed some 60,000 workers to help pay for similar deals – a figure Kraft denies.

But just a week after sealing the deal, Kraft confirmed the closure of the famous Cadbury factory at Somerdale that makes Crunchies and Curly Wurlys. During the takeover, Kraft had said it believed that it, unlike Cadbury, would be able to continue operating this factory. But in an apparent U-turn, four hundred jobs now had to go. 'This sends the worst possible message to the 6,000 other Cadbury workers in Britain,' said Unite. 'It tells them Kraft cares little for their workers.' Members of Parliament too were outspoken about what they saw as cynical manipulation. 'Kraft has treated the British Parliament with contempt,' said Liberal Democrat MP Matthew Oakshott.

The House of Commons Business, Innovation and Skills Committee investigated the closure of Somerdale, but Irene Rosenfeld failed to appear, sending Kraft's vice president for corporate and legal affairs, Marc Firestone, in her place. 'I am terribly sorry,' he said. He told the sceptical committee that Cadbury's plan to close Somerdale was more advanced than Kraft had initially appreciated, and pledged that there would be no further cuts in Cadbury's UK manufacturing for two years, although he could provide no guarantees beyond that. The MPs' report concluded that Kraft had acted 'irresponsibly and unwisely'.

The effect of the merger on the workforce is not simply worry over job security. The wider community around Bournville benefited for years from the use of chocolate wealth to fund schools and colleges, hospitals, convalescent homes, churches, housing and sporting facilities. These contributed to the local sense of unity and belonging, and also brought employment to the area. The newspaper columnist A.N. Wilson, whose father helped to create village houses for the Wedgwood workforce in Staffordshire, points out that the thriving communities created by enlightened nineteenth-century business leaders 'lie in sad contrast to the antisocial attitudes of modern business magnates who think only of profit and the shareholder'. Writing in the *Daily Mail* on 23 January 2010, he argued that the globalisation of the marketplace 'has made us all come socially adrift', adding, 'We are all victims of the "hostile takeover" of one kind or another.' Needless to say, community leaders and Birmingham MPs campaigned in Westminster against the sale of Cadbury.

The growing powerlessness of national governments in the face of these global deals is highlighted by the statements of politicians. The then Prime Minister Gordon Brown, speaking at a Downing Street press conference, declared: 'We are determined that the levels of investment that take place in Cadbury in the UK are maintained, and . . . that jobs in Cadbury can be secure.' But he had no powers to ensure that this would in fact be the case.

The role played by the Royal Bank of Scotland was seen as a bitter betrayal. The bank, which was 84 per cent owned by taxpayers after the government bailout during the credit crunch, joined the syndicate that funded Kraft, offering a £630 million loan facility. 'When British taxpayers bailed out the bank, they would never have believed that their money would be used to put British people out of work. Isn't that plain wrong?' argued the Liberal Democrat leader Nick Clegg in a stormy House of Commons debate on 20 January. City columnist Anthony Hilton pointed out, 'We have an economy dangerously skewed towards financial services, and the whole nation pays the price.' The *Guardian* summed up the anger: 'This is an old-fashioned Square Mile stitch-up, driven through by City short-termists.'

So how did it happen? For Sir Dominic Cadbury, at the heart of the issue is the changing concept of ownership inherent in our modern form of shareholder capitalism. 'There's no ownership concept,' he says – at least not in the traditional sense that his Quaker capitalist forebears understood, of stewardship and long-term planning. 'It comes back to the role of the shareholder – the shareholder is the owner of the business. But the difficulty with all this is that they are not *acting* as owners of the business. There are thousands of shareholders in Cadbury who probably did not want to sell their shares to Kraft. But they didn't have a vote, because if you are the average shareholder, you don't hold your shares personally, but through your pension scheme or your bank. In the case of Cadbury, sixty fund managers made the decision.' And fund managers are under pressure to deliver short-term performance targets, rather than to focus on long-term wealth creation.

Hedge funds demonstrate the extremes of short-termism. 'The hedge funds are "owners" whose motivation is to see that the company disappears,' says Dominic. 'By definition, they have no sense of obligation and no sense of responsibility for the company whatsoever.' By the end of the bidding process, hedge funds 'owned' more than 30 per cent of Cadbury and were happy to sell for a 20p profit – a stark contrast to the dedicated Quaker capitalist founders who nurtured the company from its humble beginnings. 'One day you had the Cadbury company, the next day you didn't,' says Dominic. 'Gone. One hundred and eighty years of history down the tube, and I would argue 180 years of being a beacon of good practice. Something very precious got lost that day. Gone. And it was so easy.'

In a keynote speech to the City on 1 March 2010, Peter Mandelson, then Secretary of State for Business, pointed out, 'It is hard to ignore the fact that the fate of a company with a long history and many tens of thousands of employees was decided by people who had not owned the company a few weeks earlier and had no intention of owning it a few weeks later.' He argued that board directors should consider the interests of all stakeholders in a business: employees, suppliers, and a company's brands and capabilities, as well as the shareholders. He urged them to act more like 'stewards' and less like 'auctioneers'.

Some MPs are calling for a 'Cadbury law' to bring about wide-ranging changes in the way foreign takeovers of British companies are managed. This would include a measure of protection for certain 'strategic' companies in the national interest. They question the government's belief in an open-door policy, as many British companies – with products including glass, steel, chemicals, a raft of public utilities and confectionery – have recently slipped into foreign ownership.

By contrast, the Swiss have always protected Nestlé, allowing their food and chocolate industries to flourish. 'In France, the loss of a "Cadbury" would have been out of the question,' says former Cadbury chairman Roger Carr. 'Germany believes that strength at home is the first step to success abroad. In Japan, selling a company over the heads of management is unthinkable. And in the United States, regulations exist to protect strategic assets.' Even Hershey has in effect been protected from foreign takeover by its relationship with the Hershey Trust.

The fate of Cadbury has also prompted a review of the City's Takeover Code. Voting rights for shares acquired during a hostile bid could be withheld to stop short-term speculation determining the fate of a company. Alternatively, greater voting rights could be given to longer-term shareholders in hostile takeovers, or the level of acceptance required for takeovers could be raised from 50.1 to 60 per cent of the share register.

While many in the City are wary of reform, not least because deals generate fees and bonuses, industry leaders have argued in favour of change. The Cadbury takeover has highlighted the extent to which shareholder ownership of many British companies has shifted in the last decade to become much more international. With old loyalties and ties broken, many see it as a priority to foster incentives for long-term share ownership. Without such changes, business managers point out, there can be a serious misalignment between the needs of companies to build long-term growth and the needs of international shareholders, who are primarily interested in short-term trading. Paradoxically, while fund managers may be judged on the quarterly returns they bring – and their bonuses may depend on them – the beneficiaries of the funds under management, such as pension funds, are dependent on successful long-term business growth.

Why has the balance tipped to short-termism? 'Greed', in the view of Todd Stitzer. 'People want money fast, and they don't really care. They just don't care what it takes to get it.' He points to the fragmentation and disconnection that has come with increased scale. The connection between the people who make the raw materials, the people who make the goods in factories and the people who finance the factories has been broken, 'so no one feels responsible for how it works'. This, he believes leads to a culture of self-interest, rather than genuine wealth creation by innovation and increased capability. What is more, he warns, the current system can distort the notion of value, and is open to abuse. Financial information about a company can be presented in a way that makes it look more valuable. Those with stock options or shares can 'manipulate the financials' to increase their value. 'The world has borne witness to legions of company directors doing this.'

For others, there is a deeper unease at the heart of our current system of shareholder capitalism, highlighted by the chocolate wars, that goes well beyond tinkering with the takeover code. A succession of smaller firms in Britain alone – Fry, Terry, Rowntree, Mackintosh, and finally Cadbury – have disappeared into two giant corporations, Nestlé and Kraft, which have annual turnovers that exceed the budget of many small countries. This in turn can mean that governments are overly influenced by the growing power of such global institutions and their shareholder lobby, and wealth and power are concentrated into fewer and less accountable hands. How democratic is this? Where does it leave the British chocolate workers, now in danger of becoming a mere commodity swelling the red column in some balance sheet in headquarters in Zürich, Vevey or Chicago? And what does the future hold for the cocoa growers, faced with ever more oligopolistic purchasers under pressure to deliver to shareholders?

It has taken 186 years to build the Cadbury company to its global position. For Sir Adrian Cadbury, one of the firm's longest-serving chairmen, the danger of Kraft taking over is that 'it is very easy for a larger firm effectively to destroy the *spirit* of the firm they take over'. The indefinable qualities that make up the character and spirit of a company, he argues, develop over many years, and depend on its history, values, international brands and the quality of the people who work for it.

Four generations ago, Adrian and Dominic's great-grandfather John Cadbury in his Victorian shop, wooden planks on the floor, tins of cocoa on display, would have struggled to comprehend that his modest enterprise, often on the verge of failure, would evolve into a global business worth billions. Would he turn in his grave at Kraft's hostile bid? I don't think so. But if, as he sat upright in his hard-backed chair, a true Puritan conceding not one inch to self-indulgence, he could hear the language of today's deal-makers, with their 'revenue synergies', 'vision into action' and 'instant consumption channels', would he recognise the spirit of what he was trying to achieve? I think not. It would be hard to see anything close to the motives that drove him, and the spirit in which he and his sons founded the business, in today's leaders. Would he lament that something was missing in the modern world? I believe he would.

The altruistic objectives of the nineteenth-century Quaker capitalists like John and George Cadbury and Joseph Rowntree, and the spirit in which they pursued them, appear as far removed from the greed of the modern corporate world – exemplified by the worst excesses of the recent financial crisis – as it is possible to be. The Kraft takeover represents a poignant symbolic end to this remarkable business enterprise that had its origins in the religious thinking of the English Civil War, and was to become integral to British culture.

Will Kraft act for the betterment of the world – not just the top management? Will it be a tangible force for good in our global village? It is difficult not to feel sceptical. And that is why, despite all the benefits of globalisation and the excitement of giant takeovers, it is hard not to believe that something irreplaceable and immeasurable in the neat columns of a balance sheet has been discarded as effortlessly as a sweet wrapper.

Epilogue

Whatever lies in store for Britain's chocolate industry, the trusts created by the pioneer chocolatiers will survive. George Cadbury's Bournville Village Trust has grown into a thriving enterprise that is still run principally by the direct descendants of George and his brother Richard. The trust is responsible for more than 8,000 properties and 1,100 acres across the West Midlands and Shropshire, as well as 2,500 acres of farmland to preserve the green belt around south-west Birmingham: a small piece of England that cannot be signed away.

Around the factory in Bournville, George Senior's utopian village has grown to 6,000 houses nestled around the original parks and playing fields. 'People still come from around the world to see Bournville,' says Duncan Cadbury, chairman of the trust's Housing Services Committee. 'They've heard it is a garden village which has worked. They even come to hear the carillon on the Junior School.' Apart from the Bournville Village Trust, the Barrow Cadbury Trust and other family benevolent trusts between them give over 250 grants a year.

Like George Cadbury's, Joseph Rowntree's voice still carries into the twenty-first century thanks to the trusts he established. His original three – the Village Trust, the Charitable Trust and the Social Service Trust – have been modified by their trustees to adapt to modern times. True to the analytical spirit of their founder, they remain heavily involved in investigating the causes of social problems. The Village Trust – now known as the Joseph Rowntree Foundation – bestows more than £10 million annually, and is one of the largest foundations in England. Its research projects cover a wide range of issues, including the causes of persistent poverty, building public support to end poverty, and – in an intriguing full circle – the effects of globalisation on poverty. The Joseph Rowntree Housing Trust, established in 1968, manages housing projects across Yorkshire, including the village of New Earswick, which has grown to 2,500 homes.

And what of Britain's once lively chocolate industry? The sweet aroma of chocolate still wafts over the suburbs of York and down Haxby Road to Joseph Rowntree's original factory, which is owned today by Nestlé. The factory is now surrounded by metal railings, and is patrolled by security guards. There is one visible reminder of the company's founder in the Joseph Rowntree Lodge – now derelict – near the main entrance. As I pressed my face to the window to glimpse inside, I saw little more than gathering dust and papers scattered on the floors of the empty rooms before a guard with a dog came to move me on.

Other centres of excellence have fared less well. Terry's gargantuan works in York, now closed down, stand as a poignant monument to past glories, the telltale broken glass in the yard a symbol of their abandonment. Could this be the fate that lies in store for Bournville under Kraft's management, I wondered.

Fry's chocolate works in Bristol suffered a more extreme fate than those of its Quaker rivals. After the business transferred to Somerdale in the 1930s, the great citadel in Union Street was severely damaged in a bombing raid in the Second World War. Today the Fry business, which once proudly claimed to be the largest chocolate company in the world, has been reduced to boxes of archives at Bournville and Somerdale, and in the Bristol Records Office.

And what of the Quaker movement that inspired these great chocolate enterprises, and proved such an astonishing force in the early industrial age? I went in search of its headquarters in Euston, in central London. Stepping inside Friends House was like entering a different world. A sudden hush prevails, creating a totally different atmosphere from the thunderous Euston Road outside. Beyond the stone-clad hall, lined with straight-backed benches Puritan in their simplicity, are corridors leading to a shady courtyard. It is easy to picture forebears and business leaders gathering here for passionate meetings to discuss the pressing issues of the day.

I met Helen Drewery, head of the international department of the Society of Friends. 'People are surprised to discover that we exist,' she told me. There was a time when one in ten people in Britain were Quakers, but today there are only 15,000 members. 'Quakers don't put very much store on dogma or Church hierarchy,' she said. 'We put our energy into trying to make the world a better place.' In the neatly kept archives was plenty of evidence to support her point: the anti-slavery movements of earlier centuries; the Kindertransports that relocated Jewish children before the Second World War; numerous famine-relief programmes; and, more recently, support for setting up the Child Poverty Action Group and Oxfam. Today Quakers have a presence in the world's major trouble spots, and there are Quaker offices in New York and Geneva that work through UN agencies. Helen Drewery described an 'accompaniment project' on the West Bank, which she calls 'a ministry of presence' to help people feel that 'the world has not forgotten about them', and to facilitate communication between opposing sides. 'People say that we punch above our weight,' she added, 'but that never feels like quite a comfortable phrase to use – not for a Quaker.'

A force for the good, surely, a quiet, sane voice that is still there for those who want to hear it in this noisy century, putting the case for ending conflict between cultures and religions, and nurturing peace. In today's material world, where the breathless glamour of celebrity culture holds the public in thrall, the Quaker message has become stifled, shut out from the boardrooms in the City of London just a mile away. It is hard to imagine today's business leaders giving more than a passing thought to the claim of George Fox that the inner light is within us all. But those nineteenth-century entrepreneurs who made this their quest did succeed for a brief period in putting the remarkable Quaker movement in the spotlight. In the process they illuminated a different work ethic, on a more human scale, between master and man.

Acknowledgements

I was fortunate to receive help from a great many people who made this account of the chocolate Quaker dynasties and their rivals possible. For expert knowledge on the chocolate branch of the Cadbury family I would particularly like to thank Sir Adrian Cadbury, one of the firm's longest-serving chairmen. It was an inspiration to discuss the history of the company and visit key locations with such a welcoming and informed guide, an experience that I will always look back on with great pleasure. Thank you, too, to former chief executive and chairman Sir Dominic Cadbury for his insights into a critical era in the firm's history from the 1960s to 2000, and for his patience in dealing with follow-up queries.

I owe a great deal to two members of my family who are no longer living. When I was young, my father, Kenneth Cadbury, and my uncle, Michael Cadbury, provided wonderful accounts of Quaker forebears in the chocolate branch of the family. Perhaps unknowingly, they each embodied different aspects of Quaker culture, and helped to inspire this account.

My cousin Duncan Cadbury provided invaluable assistance with research into Quaker history and the numerous family trusts. It was a pleasure to discuss the City perspective with Peter Cadbury, a grandson of George Junior and former deputy chairman of Morgan Grenfell. Many others in the wider Cadbury family shared compelling insights into the characters and history, which I have distilled into this account. Any views expressed in the book, however, are my own, and do not represent a unified Cadbury view.

For events at Cadbury since 2000, when family members were

no longer members of the board, I am grateful to the last chief executive, Todd Stitzer, for discussing his period of leadership and the Kraft takeover. I am also indebted to Cadbury's last chairman, Roger Carr, for his fascinating account of the takeover and the issues it raised. Cadbury's head of sustainability, David Croft, global corporate affairs director Alex Cole and others at Cadbury provided a wealth of information. Thanks are also due to Sarah Foden, the information manager, and to Jackie Jones at the Cadbury archives in Bournville. Sarah provided generous assistance with numerous queries, and gave up valuable time to comment on the manuscript.

At Kraft Foods, I am indebted to Steve Yucknut, head of sustainability, Michael Mitchell, head of external communications, and Becky Tousey, head of archives, for dealing with queries relating to the company's history and the takeover. At Nestlé, archivist Tanja Aenis in Vevey, Switzerland, and Alex Hutchinson at Nestlé's Confectionery Heritage in York investigated historical questions on the early years of Nestlé and Rowntree. Tammy Hamilton at the Hershey Community Archives provided invaluable assistance with the project. At Altria, John Marshall in corporate communications fielded queries pertaining to tobacco litigation. For insights into Mars and other chocolate companies, I received excellent support from a number of specialist archives; I have credited them in the bibliography.

At the Society of Friends, I would like to thank Helen Drewery, General Secretary of Quaker Peace and Social Witness, for her perspective on the Quaker movement. Timothy Phillips, chairman of the Quakers and Business Group, raised key questions about changing ethical values. In the library of the Society of Friends, I am indebted to Josef Keith, Joanna Clark, Tabitha Driver, Beverley Kemp, Jennifer Milligan and Julia Hudson for responding to queries over many months, and I was particularly touched when they traced the unpublished 'family book' written by Richard Cadbury in the 1860s, which provides a vivid insight into the early years.

There are so many others who helped during the months of research that it would be impossible to name them all, but I would like to thank John Crosfield; Steven Burkeman, chairman of the Rowntree Society; Timothy Newman at the International Labour Rights Forum in Washington, DC; Dave Goodyear at Fairtrade; Hugh Evans for discussing aspects of Fry history; Adam Leyland, editor of the *Grocer*; and John Bradley and Martin Simons.

At Harper Press it was a pleasure to work with Martin Redfern, and I have appreciated his skilled editorial advice at each stage in the production of the manuscript. I am also grateful to Robert Lacey for his thoughtful comments on the text and to Sarah Hopper for her picture research. At Curtis Brown, Gordon Wise's insights and encouragement on the project over many months proved invaluable.

Lastly, I would like to express my thanks to Julia Lilley, without whom this book would not have been written. Her generous support for and great faith in the project have kept me going through the long months of writing.

Bibliography

Archives

From the Library of the Society of Friends, Friends House, London:

Cadbury, Richard. 'Family Book', Temp MSS 996, 1863
Fry, Joseph. Records 1727–1787, MS Vol. S272
Fry, Joseph Storrs. Correspondence, MS Box 16/4/28
Fry, Joseph Storrs. Correspondence, Temp MSS 668/1
Successive editions of the Quaker *Book of Discipline*:
Christian and Brotherly Advices, manuscript version, 1738–1749
Extracts from the Minutes and Advices of the Yearly Meeting of Friends 1783
Rules of Discipline of the Religious Society of Friends, with Advices, 1834
Extracts from the Minutes and Advices of the Yearly Meeting held in London Relating to Christian Doctrine, Practice and Discipline, 1861

From the Cadbury Archives, Bournville, Birmingham:

Bournville: A Descriptive Account of Cocoa and its Manufacture. 1880
Bournville Works magazine, 1900–1970
Bridge Street Album. 1830–1900. 010 003250 AZ1
Cadbury and Fry company publications
Cadbury Brothers Fancy Boxes and artwork
Cadbury Brothers overseas records. 030–090
Fry, J.S. & Sons. *Bicentennial Issue: Fry's Works Magazine 1728–1928*. Broadmead, Bristol: Partridge & Love, 1928
Fry Company papers. 910–918.2
Fry's Works magazine, 1922–1929. 910.2 001687
Industrial Record: A Review of the Interwar Years, 1919–1939. Cadbury Bros/ Pitman, 1945
Somerdale magazine, 1961–1968. 910.2 001696
Thirty Years of Progress: A Review of the Growth of the Bournville Works. 1910

From the Cadbury Papers Collection, Birmingham City Archives, Central Library, Birmingham:

Cadbury, Barrow. Letters and notes/gift from employees. UBL MS 466/211–221
Cadbury, Benjamin Head, Richard Tapper, and others. Family letters, vols 1–3. UBL MS 466/85 1
Cadbury Brothers. Bull Street and Bournville early images. UBL MS 466/41/1/2/3/8/
Cadbury Brothers. Business Affairs: Ledgers and Accounts. UBL MS 466A/ 39
Cadbury Brothers. Business Affairs: Wages from 1859–1864. UBL MS 466A/ 1–10
Cadbury Brothers. UBL MS 466A/163–165
Cadbury Brothers. 16 December. UBL 1910 180/2
Cadbury, Candia. Letters to her sons John and George, including 'A Mother's Affectionate Desire for Her Children'. UBL MS 466/971–53; 466/98; 466/99/1–8; 466/100/1–2
Cadbury, Elizabeth. Journal. UBL MS 466/431–434
Cadbury, George. Letters. UBL MS 466/209/1–13 and 209/219
Cadbury, William. Correspondence. November 1907. UBL 1907: 180/365
Cadbury, William. Miscellaneous personal album. UBL MS 466/32
Cadbury, William. UBL 1903 180/336
Notebook. UBL 1909 299–300

From the Rowntree Papers Collection, Borthwick Institute of Historical Research, University of York:

Information on Tanners Moat site in the 1890s, including insurance documents, plans for new Haxby Road site, and other documents relating to Rowntree's factories. HIR 2 10–12 Archive 398
Reports on the van Houten process. HIR1/5–13 & HIR 2 1–7
Rowntree, Joseph. Personal notebooks, correspondence, business memoranda, drawings, etc. Ledger HIR/1 Sections 1–4
Rowntree, Joseph. Undated agreements on work, wages, machinery, etc. HIR 2 8–12 Archive 376

Manuscripts

Fry, Theodore. 'A Brief Memoir of Francis Fry'. Unpublished typescript. British Library, London: 1887
_____. History of J.S. Fry and Sons. Unpublished typescript
The London Commodity Exchange. Unpublished typescript from the London Stock Exchange, 1961. City Library, London: FO PAM 1486
Strong, L.A.G. The Story of Rowntree. Unpublished typescript. 1948

Wallace, Paul. The Wallace manuscripts, typescript for the Hershey Trust, 1955

Books

Acheson, T.W. 'The National Policy and the Industrialisation of the Maritimes'. In *Canada's Age of Industry, 1849–1896*, edited by M.S. Cross and G.S. Kealey, 62–94. Toronto: McClelland & Stewart, 1982

Alexander, Helen. *Richard Cadbury of Birmingham*. London: Hodder & Stoughton, 1906

Barclay, H.F., and A. Wilson-Fox. *A History of the Barclay Family*. 1934

Barringer, E.E. *Sweet Success: The Story of Cadbury and Hudson in New Zealand*. Dunedin Cadbury Confectionery Ltd, 2000

Bartlett, Percy W. *Barrow Cadbury: A Memoir*. London: Bannisdale Press, 1960

Beable, W. H. *Romance of Great Businesses*. London, 1926

Beckett, Stephen T. *The Science of Chocolate*. Cambridge: Royal Society of Chemistry Publishing, 2008

Benson, S.H. *Wisdom in Advertising*. London: John Murray, 1901

Bradley, John. *Cadbury's Purple Reign: Chocolate's Best-loved Brand*. Chichester: John Wiley & Sons, 2008

Braithwaite, William C. *The Beginnings of Quakerism*. London: Macmillan, 1912
_____. *The Second Period of Quakerism*. York: Sessions of York, 1979

Brayshaw, A. Neave. *The Quakers: Their Story and Message*. Harrogate: Robert Davis, 1921

Brenner, Joël Glenn. *The Emperors of Chocolate: Inside the Secret Worlds of Hershey and Mars*. New York: Random House, 1999

Briggs, Asa. *Social Thought and Social Action: A Study of the Work of Seebohm Rowntree, 1871–1954*. London: Longmans, Green & Co., 1961
_____. *Victorian Cities*. London: Penguin, 1963

Broekel, Ray. *The Great American Candy Bar Book*. Boston: Houghton Mifflin, 1982
_____. *The Chocolate Chronicles*. Wallace Homestead Book Co., 1985

Cadbury, Christabel. *Robert Barclay: His Life and Work*. London: Headley Bros, 1912

Cadbury, Edward. *Experiments in Industrial Organisation*. London: Longmans, Green & Co., 1912

Cadbury, Edward, and George Shann. *Sweating*. London: Headley Bros, 1907

Cadbury, Edward, George Shann and Cecil Matheson. *Women's Work and Wages*. London: T. Fisher Unwin, 1906

Cadbury, George. *Conurbation: A Survey of Birmingham and the Black Country*. Birmingham: West Midland Group, 1948

Cadbury, Richard. ['Historicus', pseudo.]. *Cocoa: All About It*. Birmingham: Sampson, Low & Marston, 1892

Cadbury, William. *Labour in Portuguese West Africa*. London: Routledge, 1910

Cadbury Brothers. *Cocoa and its Manufacture.* Carlisle: Hudson Scott & Sons, 1880

_____. *Industrial Challenge: The Experience of Cadburys of Bournville in the Post-War Years.* London: Pitman Publishing, 1964

Carnegie, Andrew. *The Gospel of Wealth.* London: Penguin, 2006

Carr, David. *Candy Making in Canada.* Toronto: The Dundurn Group, 2003

Chernow, Ron. *Titan: The Life of John D. Rockefeller.* New York: Vintage Books, 1999

Chinn, Carl. *The Cadbury Story: A Short History.* Studley, Warwickshire: Brewin Books, 1988

Church, Roy A. *The Great Victorian Boom: 1850–1873.* London: Macmillan, 1896

Clarence-Smith, William Gervase. *Cocoa and Chocolate: 1765–1914.* London: Routledge, 2000

Coady, Chantal. *Chocolate: The Food of the Gods.* San Francisco: Chronicle Books, 1993

Coe, Sophie D., and Michael Coe. *The True History of Chocolate.* London: Thames & Hudson, 1996

Cook, L. Russell. *Chocolate Production and Use.* New York: Books for Industry, 1972

Crosfield, J.F. *A History of the Cadbury Family.* 2 vols. Cambridge: Cambridge University Press, 1985

Crutchley, Geo. W. *John Mackintosh: A Biography.* London: The National Sunday School Union, 1921

D'Antonio, Michael. *Hershey: Milton S. Hershey's Extraordinary Life of Wealth, Empire and Utopian Dreams.* New York: Simon & Schuster, 2006

Diaper, Stefanie. 'J.S. Fry and Sons: Growth and Decline in the Chocolate Industry, 1753–1918'. In *Studies in the Business History of Bristol,* edited by Charles Harvey and Jon Press, 33–53. Bristol: Bristol Academic Press, 1988

Duffy, James. *A Question of Slavery.* Cambridge, MA: Harvard University Press, 1967

Emden, Paul H. *Quakers in Commerce.* London: Sampson Low, Marston & Co., 1940

Fergusson, Niall. *Empire: How Britain Made the Modern World.* London: Penguin, 2003

Finch, R. *A World Wide Business.* Birmingham: n.d.

Fitzgerald, Robert. *Rowntree and the Marketing Revolution: 1862–1969.* Cambridge: Cambridge University Press, 1995

Folster, David. *Ganong: A Sweet History of Chocolate.* New Brunswick, Canada: Goose Lane Editions, 2006

Fraser, W.H. *The Coming of the Mass Market: 1850–1914.* London: Macmillan, 1981

Fry, J.S. & Sons. *Fry's of Bristol Established 1728.* Bristol, n.d.

Fry, J.S. & Sons, and Cadbury Bros Ltd. *The British Cocoa and Chocolate Co. Ltd.* 1948

Gardiner, Alfred G. *Life of George Cadbury.* London: Cassell & Co., 1923

Grivetti, Louis E., and Howard Shapiro. *Chocolate: History, Culture, and Heritage.* Hoboken, NJ: John Wiley & Sons, 2009

Harris, J.H. *Dawn in Darkest Africa.* London: Smith, Elder & Co., 1912

Harwich, N. *Histoire du Chocolat.* Paris: Editions Desjonqueres, 1992

Head, B. *The Food of the Gods: A Popular Account of Cocoa.* London: Routledge, 1903

Heer, J. *World Events 1866–1966: The First Hundred Years of Nestlé.* Lausanne, Switzerland: Imprimeries Réunies, 1966

Hewett, C. *Chocolate and Cocoa: Its Growth and Culture, Manufacture and Modes of Preparation for the Table.* London: Simpkin, Marshall & Co., 1862

Hinkle, Samuel F. *Hershey: Far Sighted Confectioner, Famous Chocolate, Fine Community.* New York: Newcomen Society, 1964

Hobsbawm, Eric. *The Age of Capital: 1848–1875.* London: Weidenfeld & Nicolson, 1975

Hodgkin, J.E., ed. *Quakerism and Industry [Record of the Conference of Employers at Woodbrooke April 1918].* London, 1925

Howard, Ebenezer. *Tomorrow: A Peaceful Path to Real Reform.* London: Routledge, 1898

Knapp, Arthur W. *Cocoa and Chocolate: Their History from Plantation to Consumer.* London: Chapman & Hall, 1920

———. *The Cocoa and the Chocolate Industry: The Tree, the Bean, the Beverage.* London: Pitman Publishing, 1923

Markham, Leonard. *York: A City Revealed.* Stroud, Gloucestershire: Sutton, 2006

Marks, W. *George Cadbury Junior.* Birmingham: n.d.

Mathias, Peter. *The First Industrial Nation: The Economic History of Britain, 1700–1914.* London: Methuen, 1969

Milligan, Edward H. *Biographical Dictionary of British Quakers in Commerce and Industry.* York: Sessions Book Trust, 2007

Murphy, Joe. *New Earswick: A Pictorial History.* York: Sessions Book Trust, 1987

———. *The History of Rowntree's in Old Photographs.* York: York Publishing Services, 2007

Nevinson, Henry W. *A Modern Slavery.* New York: Schocken, 1968. First published 1906 by Harper & Bros

Nickalls, John L., ed. *The Journal of George Fox.* Philadelphia, PA: Religious Society of Friends, 1997

Othnick, J. 'The Cocoa and Chocolate Industry in the Nineteenth Century'. In *The Making of the Modern British Diet,* edited by Derek T. Oddy and Derek Miller, 77–90. London: Croom Helm, 1976

Pfiffer, Albert. *Henri Nestlé: 1814–1890.* Vevey, Switzerland: The Nestlé Corporation, 1995

Richardson, Paul. *Indulgence: Around the World in Search of Chocolate.* London: Little, Brown, 2003

Richardson, Tim. *Sweets: A History of Temptation.* London: Bloomsbury, 2002

Roberts, Jane S. *Drink, Temperance and the Working Class in Nineteenth-Century Germany.* Boston: Allen & Unwin, 1984

Rogers, T.B. *A Century of Progress: 1831–1931.* Birmingham: Cadbury Bros, 1931

Rosenblum, Mort. *Chocolate: The Bittersweet Saga of Dark and Light.* New York: North Point Press, 2006

Rowntree, Benjamin Seebohm. *Poverty: A Study of Town Life.* London: Thomas Nelson & Sons, 1901

_____. *The Human Needs of Labour: Land and Labour.* London: Thomas Nelson & Sons, 1910

Rowntree, C. Brightwen. *The Rowntrees of Riseborough.* York: Ebor Press, 1989

_____. *The Way to Industrial Peace and the Problem of Unemployment.* London, 1914

Rowntree, John Stephenson. *A Memoir of Joseph Rowntree, 1801–89.* Birmingham: privately printed

Rowntree, Joseph. *Pauperism in England and Wales.* 1865

Rowntree, Joseph, and Arthur Sherwell. *The Temperance Problem and Social Reform.* London: Hodder & Stoughton, 1899

Rowntree & Co. *Industrial Betterment at Cocoa Works.* York: 1905, 1910, 1914

Rowntree & Son. *A Century and a Half of Progress.* 1930

Satre, Lowell J. *Chocolate on Trial: Slavery, Politics and the Ethics of Business.* Athens, GA: Ohio University Press, 2005

Schwarz, Friedhelm. *Nestlé: The Secrets of Food, Trust and Globalisation.* Ontario, Canada: Key Porter Books, 2002

Sharman, Cecil. *George Fox and the Quakers.* Richmond, IN: Friends United Press, 1991

Shippen, Katherine, and Paul Wallace. *Milton S. Hershey.* New York: Random House, 1959

Smith, Page. *The Rise of Industrial America.* New Haven, CT: Yale University Press, 1986

Snavely, Joseph. *An Intimate Story of M.S. Hershey.* Hershey, PA: privately printed, 1957

Sprüngli, Rudolph R. *150 Years of Delight: Chocoladefabriken Lindt & Sprüngli 1845–1995.* Switzerland, 1995

Stranz, Walter. *George Cadbury: An Illustrated Life.* Aylesbury: Shire Publications, 1973

Taylor, Alan J. *Progress and Poverty in Britain: 1780–1850.* London: Harper

Teiser, R. *An Account of Domingo Ghirardelli and the Early Years of the D. Ghirardelli Company.* San Francisco: D. Ghirardelli Co., 1945

Terrio, Susan J. *Crafting the Culture and History of French Chocolate.* Los

Angeles: University of California Press, 2000

Terry, J. & Sons. *Terry's of York: 1767–1967*. Privately printed by Newman Neame

Townsend, Richard F. *The Aztecs*. London and New York: Thames & Hudson, 1992

Turner, Ernest S. *The Shocking History of Advertising*. London: Michael Joseph, 1952

Urquhart, D.H. *Cocoa*. London: Longmans, Green & Co., 1955

Vernon, Anne. *A Quaker Business Man: The Life of Joseph Rowntree 1836–1925*. London: Allen & Unwin

Wagner, Gillian. *The Chocolate Conscience*. London: Chatto & Windus, 1987

Walvin, James. *The Quakers: Money and Morals*. London: John Murray, 1997

Whitney, Janet. *Elizabeth Fry: Quaker Heroine*. London: George Harrap & Co., 1937

Wild, Anthony. *Black Gold: A Dark History of Coffee*. London: Harper Perennial, 2005

Williams, C.T. *Chocolate and Confectionery*. London: L. Hill, 1953

Williams, Iolo A. *The Firm of Cadbury: 1831–1931*. London: Constable & Robinson, 1931

Wood, Stephen. *A History of London*. London: Macmillan, 1998

Woolf, Virginia. *Roger Fry: A Biography*. London: The Hogarth Press, 1940

Worstenholm, Luther. 'Joseph Rowntree: 1836–1925'. A typescript memoir and related papers. York: Sessions York, 1986

Articles

It would not be possible to list all articles consulted. This is a guide to the key articles:

Banks, Myron. 'Mars to Expand Factory'. *Chicago Daily Tribune*, 6 April 1958

Burtt, Joseph. 'How America Can Free the Portuguese Cocoa Slave'. *Leslie's Illustrated Weekly*, 14 October 1909, 368–9

_____. 'My Success in America'. *Leslie's Illustrated Weekly*, 16 December 1909, 608

Chase, Al. 'Standard Set by Mars Plant Built in 1928'. *Chicago Daily Tribune*, 15 November 1953

Daily Mail (on outbreak of Boer War), 11 October 1899

Dombrowski, Louis. 'Candy Makers Unspoiled by Sweet Smell of Success'. *Chicago Daily Tribune*, 5 January 1961

Elwood, Berman. 'Mars Embattled over Succession'. *Chicago Daily Tribune*, 10 June 1959

'The Factory in a Garden'. *Cosmopolitan*, June 1903

Ferguson, Richard. 'At Mars, Sweet Success'. *The Times*, 8 May 1953

Gross, Alan. 'Sweet Home Chicago'. *Chicago* magazine, February 1988

Gussow, Don. 'Forrest Mars'. *Candy Industry and Confectioners Journal*, 1966

Hobhouse, Emily. Letter to the editor. *The Times*, 27 June 1901

Kessler, Ronald. 'Candy from Strangers'. *Regardie's* magazine, August 1986
Lippman, Thomas W. 'The Mars Empire: How Sweet It Is'. *Washington Post,* 6 and 7 December 1981
Nevinson, Henry. 'The New Slave Trade'. *Harper's Monthly Magazine,* August 1905–February 1906
Poe, Tracy. 'Sweet Home Chicago: Candy Makers Made City Their Capital'. *Chicago Tribune,* 16 July 1997
Saporito, Bill. 'Uncovering Mars's Unknown Empire'. *Fortune,* 26 September 1988
_____. 'The Eclipse of Mars'. *Fortune,* 28 November 1994
'The Sweet, Secret World of Forrest Mars'. *Fortune,* May 1967
'Where Happiness and Health Go Hand in Hand with a Great Enterprise'. *Business World,* June 1903
Young, James C. 'Hershey Unique Philanthropist'. *New York Times,* 18 November 1923

Index